ECOLOGY
OF
RIVER GANGA

ECOLOGY OF RIVER GANGA

D.R. Khanna
Gagan Matta

BIOTECH
2011
BIOTECH BOOKS

ISBN 978-81-7622-225-9

Published by

BIOTECH BOOKS

4762-63/23, Ansari Road, Darya Ganj,
New Delhi-110002 (India)
Phone: 23262132
E-mail: biotechbooks@yahoo.co.in

Laser Typesetting by : Deep Graphics, Delhi

PRINTED IN INDIA

CONTENTS

PREFACE

Fresh water is the basic need for survival or earth. This book has been written for the students and researchers working in the field of aquatic ecology of freshwater ecosystems. Book is aiming to provide understanding of freshwater ecology of River Ganga.

With the end of the first decade of 21st century many rivers has already been in critical condition and need special conservation and preservation in country. The problem was not just started and recognized recently but it is been started since the last quarter of 20th century. In last decade the problem of freshwater and decline of river ecosystems has gown largely for which many programs, projects were also initiated.

The current book describes the detailed account of aquatic ecology of River Ganga in Uttarakhand region. The book consists of 7 chapters including bibliography. Chapters includes introduction, review cited for the current study conducted, methodology used for analyzing various parameters. The detailed account of observation with tables and graphical presentation has been given flowing with the discussion chapter in and summary of the complete study.

With keeping the importance of River Ganga in mind, the study was conducted and converted in form of books to help in future projects and researches for the sake of rivers and environment.

The authors are grateful and thankful to publishers M/S Biotech Books, New Delhi for bringing up the book in an attractive way in short span of time.

CHAPTER-1

INTRODUCTION

River Ganga, the holiest river of all rivers and lifeline of north India originates from Gangotri glaciers. It emerges after confluence of main streams of the two important rivers of hills river Bhagirathi and river Alaknanda at Devprayag. From Haridwar it flows down towards south and then south-east touching many important cities like Garh Mukteshwar, Anupshahar and Narora in Bulandshahar, metro towns like Kanpur, Allahabad, Varanasi and lastly terminates in the Bay of Bengal covering about 2,506 kms. in India.

In a report on the 'State of the planet' by UN and the Washington Bank, fresh waters are identified to be the most threatened out of all the ecosystems in the world. The fresh water consumption has outpaced global population growth by 100% during this century. An estimated 54 % of the world's fresh water resources are reported to be in use presently. The heavy load of exploitation has given rise to a long list of challenging problems, especially in Asian countries. The scarcity, water pollution, eco-shift, ecological disbalance and extinction of biological lives have an ultimate effect on the health of man and its environment. Till today, the major basins like Indus, Ganga and the Brahmaputra serve as the Water Towers of the Himalayas. These directly or indirectly get connected by the other tributaries and also large rivers. In practical terms, we all are aware of the fact that without water there is no life, so rivers are the only life giving support. They are the backbone for sustaining life and its development. It is used for drinking, washing, irrigation and also for energy generation. The rivers have their own purifying power that is called the putrification but extensive use, misuse and over exploitation of this resource as any other natural resource has reduced this natural capacity of self-cleaning of the rivers. Water is infact one of the primary requirement of both living and non-living cycles in and around us. Water quality index has been considered as one of the most effective ways to inform our policy makers and the general public about the quality trend of water.

In chemical terms it is the mono-oxide of hydrogen. Water dissociates into ($HOH \rightarrow H^{+} + OH^{-}$) due to its dielectric constants, salts get easily dissolved in water than in any other solvents. Thus water is also known as Universal Solvent. The natural water contains many minerals and salts dissolved in it. On the earth only 3% of the total water is reported to be fresh water and suitable for human consumption. Of this less than 0.01% is associated with rivers and streams. In India, 14 major river systems shares about 83% of the drainage basin. In other words, Indian rivers are supposed to carry about 16,45,000 million meters cube of water annually (Rao, 1975). The Himalayas are the source of fresh water so much so that they can be called "the cradle of major rivers" that supports the life of many ecosystems. The Himalayan rivers receive about 20-30 % of their water from melted snow during the summer and monsoon seasons. As reported by Deming (1957), if the total ice cover of the world were to melt slowly, all the rivers of the World would keep flowing for 830 years. But the degrading environment and the devastating climatic changes due to ozone hole, greenhouse effect and global warming etc. have endangered the very existence of life.

The river Ganges is regarded as the most holy and sacred rivers of the World from time immemorial. It occupies a unique position in the ethos of people of India. The river's name in Hindi, the chief Indian language, is Ganga. It is rather regarded as the river from heavens and has a special sentimental place among its worshipers. According to legend, Prince Bhagirath performed penance here to salvage the souls of his ancestors who had perished due to Sage Kapila's curse. The penance was answered and the river Ganga trickled forth from Lord Shiva's locks and its bountiful water revived the sons of King Sagara. In the tradition of Bhagirath, devout Hindus stand in the sacred waters here, praying for the salvation of their ancestors.

The *Mother river of India-'The Ganges'* has its source in an ice cave 3,139 metres above sea level in the Himalaya of northern India. The river flows toward the southeast and through Bangladesh for 2,480 kilometres to empty into the Bay of Bengal The total stretch holds a large bio-diversity in and around it self. Several tributary rivers, including the Jamuna, Ramganga, Gomati, Ghaghra, Son, and Sapt Kosi add to the water of the Ganges. The Brahmaputra River joins some of the branches of the Ganges near its mouth, and together the two rivers form a large delta.

The Alaknanda and the Bhagirathi gain the prestige of being the parent stream of this holy river. During its long course it embraces many small torrents and tributaries of varied origin. The Ganga also receives a part of its water from another source that is monsoon between July and October. The Ganga basin is known to be one of the most fertile and densely populated regions of the World. Its valley is fertile and densely populated. The river is an important trade area. The cities such as Haridwar, Varanasi and Allahabad touches the bank of this holy river and gain the privilege of hosting many sacred ceremonies and festivals right from the birth of a child to its death.

Besides being an important pilgrim centre of the Hindus, Haridwar is a major adventure sports circuit offering possibilities for river rafting, tracking and mountaineering. Haridwar is also known the World over for its herbal remedies and Ayurveda (the ancient Indian science of medicine).

The sick and crippled come hoping that the touch of the water will cure their ailments. Others come to die in the river, for the Hindus believe that those who die in the Ganges will be carried to Paradise. The World's biggest mela also called the Kumbh Mela is held at the bank of this river Lakhs of people take holy bath at this occasion with devotion faith and firm belief in it. There are many ghats and kunds as bathing sites for the pilgrims-The Teerath Yatrees. Among these Har-ki-Pauri is the center of activity in Haridwar. Before a few decades Haridwar and Rishikesh were very peaceful but with time the progressing growth in the population, unchecked immigration, industrialization and unsustained development, the quality of Ganga water at this long stretch is facing a risk of deterioration especially at sites that are near populated areas.

The Ganga cuts across the Shivalik hills and for the first time it enters the plain of the Uttrakhand state in India. Ganga river after coming from hills, reaches first time in Rishikesh and Haridwar. These cities are of great importance for Hindu pilgrims. Due to rapid industrialization, urbanization and increasing population day-by-day, water consumption rate has increased and major causes of water pollution are extended as a result of treated and untreated sewage and industrial effluents discharging into water sources. Global water consumption has rose six folds between 1900-1995. According to the United Nations, New York one third of the worlds population live in countries dealing with moderate to high water resources, streams and warns that the situation will get worse without major changes in the way the water is distributed and used. Better management of water resources is the key to mitigating water scarcities in the future and avoiding further damage to aquatic ecosystem. On the banks of river Ganga several towns/major cities and industries are situated which have no proper management of sewage drains and effluent treatment plant for controlling industrial pollution. Consequently sewage, treated and untreated industrial effluents are being discharged directly or indirectly into the river and deteriorating the water quality of river Ganga day-by-day.

It is well known that Ganga is one of the most important rivers of India and has served as cradle for Indian civilization. Although there are several large cities all along the banks of Ganga river and serves as source of water supply to these towns. Over the years the river has been indiscriminately polluted and misused. Despite of its extra ordinary resilience and recuperative capacity is severely polluted. Due to increase of population and industrialization, the water quality of river Ganga has been damaged from domestic sewage and industrial effluents that contain large number of chemicals and heavy metals. Waste materials react with each other as a result the water is polluted and may become toxic which ultimately makes the water unpotable and also severely affect the bio-productivity of the aquatic system.

Other than the daily influent of wastes, human and animal carracases, flowers, papers, clothes and plastic bags etc., and the wastewater with various characteristics from industries, domestic places and recreational sources find their ultimate way to the river. The bathing festivals like Makarshankranti, Guru Purnima, Ganga Dushera, Vaishakhi and the Kumbh mela are the high times when thousands and lakhs of pilgrims or teerath yatrees take bath daily. All these have catalyzed the chemical as well as the biological problems.

Instead of the efforts of the non-governmental and governmental agencies/organizations the indiscriminate discharge of the industrial, domestic sewage and other wastes from market, business centers, hotels and recreational centers are a continuous practice. This discharge has disturbed the aquatic environment to such an extent that it has hampered the very progress of the project of preserving the natural environmental quality of the river i.e. The Ganges. The rich macro and micro biota existing in its ecosystem is facing a severe stress, which may result in shifting and even extinction.

The quality/potability of water that is consumed defines the base line of protection against many diseases and infections. Prevention at the source i.e. end of the pipe is almost invariably the best method of ensuring safe drinking water and to be preferred to treat a contaminated water supply to render its suitability for consumption (WHO, 1985).

The effluent from various sources contains huge amount of many organic and in-organic chemicals along with the heavy metals. This has a devastating toxic effect both individually and synergistically. Not only do these alter the physico-chemical characteristics, the decrease in the dissolved oxygen (DO), increase in bio-chemical oxygen demand (BOD) and chemical oxygen demand (COD), alkalinity, other organic constituents and quantity of heavy metals but also affect the biological life prevailing in that aquatic environment. The biota present and its density are given significant importance in any reservoir or limnological study. Temperature, dissolved oxygen (DO) and organic matter are some of the important factors that control the growth of aquatic life. The studies performed in the past reflect the presence of *Escherichia coli, Salmonella sp.* and many other such pathogenic bacteria which are prominently found in fecal matter to be also present in the water of Ganga Bonde (1977). The pollution of river also creates environmental stress on the indicator bacteria in the natural water Bissonette *et al.* (1975).

Phytoplankton is supposed to be the center of aquatic food web. These serve as the food for both upper & lower strata in the food triangle. Changes in the zooplankton community structure and the predators that feed on zooplankton often add due to anthropogenic acidification of lakes. Zooplankton is microscopic free-swimming animalcule of any aquatic system. These are also the primary consumers of the phytoplankton. These serve as the major food for the fishes.

Thus, the pollution of aquatic system makes a noticeable change in its physico-chemical characteristics to an extent that the growth of the fish community is affected. The unavailability of the fish-food and blocking of migrational path are also destroying the fish breeding grounds.

This work is presentation of the studies, performed on the physico-chemical, microbiological & biological aspects of river Ganga from Rishikesh to Haridwar.

CHAPTER - 2

REVIEW OF LITERATURE

The people of ancient times seemed to be concerned about the water quality, as they had realized that it was one of the most absolute necessities for life. Different workers from India and aboard have discussed various aspects of physico-chemical and biological characteristics of water bodies from time-to-time. However, it is hard to cite the name of all scientist, limnologists and hydrobiologists who worked on the varies aspects of physico-chemical and biological characteristics of water ecology. This review includes some of the important contributions made from various parts of the globe.

Several studies were carried out on the physico-chemical properties of water. Needham and Needham (1972) presented a guide to fresh water biology. Various studies has been made by Ray *et al.* (1966), Badola and Singh (1981), Dobriyal and Singh (1981), Singh *et al.* (1982), Nautiyal and Lal (1982 a,b), Dobriyal *et al.* (1983) , Pokhriyal *et al.* (1983) & Bilgrami and Duttamunshi (1985) on the river Ganga; Dobriyal (1985), Sangu and Sharma (1985) made studies on river Yamuna and Bisht and Das (1985) presented study on abiotic parameters and aquatic insects of two cold water lakes in Kumaon. Kudesia and Verma (1985) performed a study of industrial pollution on Kali river. Nautiyal *et al.* (1986) studies the quality and potability of water of Alaknanda river and gave a note on bacterial density as an indicator of pollution Dobriyal and Singh (1987) gave a study on Rhitho plankton in Garhwal hill streams. Chopra *et.al.* (1994), studied the effect of domestic sewage on self-putrification of Ganga water at Rishikesh. Hamilton (1822) made an account of the fishes found in the river Ganga and its branches. Forel (1892) can be undoubtedly considered as the founder of modern Limnology. He was the first to study the chemical, physical and biological characteristics of the lakes. Hora (1921) made his study on some new or rare species of fishes from Eastern Himalayas.

The studies on river system continued, Velz (1947) explains the process of recovery and self-purification of river water. Abdin (1948 a, b) studied the relationship of physico-chemical factors to algal growth and seasonal distribution of plankton and sessile algae in river Nile. Chacko and Ganapati (1949) made hydrobiological observations on the river Adyar. Furthermore, Chako and Ganapati (1952) also studied the hydrobiology of Surli river of Madurai District to determine its suitability from the introduction of the rainbow fish trout. Ahmed (1952) in Pakistan reported various stocking combinations of local major carps based on the ecological conditions of pond water. Berner (1951) investigated the limnology of lower Missouri river Das and Moitra (1955) said that fish belonging to surface, column and bottom feeding categories have been sub-grouped according to food character they consume, into herbivorous, carnivorous and omnivorous. Das and Moitra (1955) studied the feeding habits of fresh water fishes of Uttar Pradesh. Griffith (1955) analyzed the plankton yield in relation to certain physico-chemical parameters in lake Michigan. Motwani *et al.* (1956) studied about the river Sone. Hora (1957) made contribution in the study of physiology of fishes. Oliff (1959) analyzed the hydrobiology of Tugela River. Holden and Green (1960) studied the hydrology of plankton of river Sokoto. Mahajan (1961) studied the fish fauna of Muzzaffarnagar district, Uttar Pradesh. The contribution of Lloyd (1961) was made by his study on the toxicity of mixture of zinc and copper sulphates to rainbow trout (*Salmo gaidnerii*). Reid (1961) investigated the ecology of inland water and estuaries at New York. Chugunova (1963) studied the growth and age in fishes.

Ward and Seibert (1963) studied the limnology of Dow Lake. Weible *et al.* (1964) declared that urban land run off also contribute to pollution of river water. Deshmukh *et al.* (1964) made studies on the water of river Kanhan (Nagpur) Ray *et.al.* (1966) studied some aspects of the river Ganga and Jamuna at Allahabad (U.P.). Mills *et al.* (1966) observed the effect of human activities on fish of river Illinois river. Ward and Wipple (1959) made further contributions in the study of the fresh water biology. Train (1967) evaluated the quality criteria for water. Venkateswarlu and Jayanti (1968) studied the hydrobiology of the river Sabarmati to evaluate the water quality. Ghosh and Basu (1968) while working in Hooghly estuary reported that the waste discharges from a chemical factory contains both inorganic and organic pollutants, which highly effects the downstream of the discharge point as compared to upstream point. Due to high dilution and heavy mixing the estuary recovers from the hazards of pollution, as it travels at a distance of about 3.5 km downstream. Muller and Forstner (1968b) studied the general relationship between the water discharge and suspended sediment concentration. Verma and Shukla (1969) observed the pollution in perennial stream Khala, by the sugar factory effluent near Laksar (Dist. Saharanpur). Das and Akhtar (1970) studied the microorganisms of Dal Lake, Kashmir. Hynes (1970) also observed the biology of polluted water. Tan (1971) studied the occurrence and ecological adaptation of zooplankton in Tansh river of Taiwan. Needham and Needham (1972) had given a guideline for study of fresh water biology Grove (1972) investigated some south-west African rivers with respect to dissolved and suspended load. Prescott (1972), Hart and Fuller (1974), Macan (1974) were among those who gave the main method for the analysis of water samples for physical, chemical and biological parameters. Cosy and Newton (1973) studied the chemical compounds and flow of the river Fromes and its tributaries. Stumm and Bilinski (1973) studied trace metals in natural water. Sehgal (1973) studied fishes and fisheries of Aligarh with notes on fish catches.

Badola & Pant (1973) made a study on the fauna of the Garhwal Hills. Jhingran (1975) made study on the fish and fisheries of India. Verma and Dalela (1975a) performed study on river pollution. Robinson and Kallar (1976) compared the water characteristics of four major northern west rivers of Virginia Dabron (1976) observed physical and chemical factors of a vernal temporary pond in western Canada. Bowakar (1976) investigated the limnology of Mwenda Mouth Lake, Korea. Agarwal *et al.* (1976) monitored the physico-chemical characteristics of river Ganga at Varanasi. Bonde (1977) observed *Eschereihia coli* as bacterial indicator of water pollution. He further reported *Salmonella* sp. and other pathogenic bacteria found in fresh water. Goyal *et al.* (1977) also noted occurrence and distribution of bacterial indicators and pathogens in canal communities. Cook (1977) monitored the environmental pollution by heavy metals. Patrick (1978) used algae as indicators of pollution to observe trace metals in water body. Jhingran and Sehgal (1978) Studied the cold water fisheries of India. Venkateshwarlu and Menon (1979) gave a list of fishes of river Ganga and its branches. Badola (1979) pioneered limnological studies on rivers of Garhwal Himalayas. Govindan and Sundaresan (1979) studied the seasonal succession of algae in relation to pollution levels in Adyar river. Bharti *et al.* (1979) noticed physiological imbalance in fresh water algae due to hexavalent chromium metal. Bilgrami and Dutta Munshi (1979) studied the limnology of river Ganga in relation to human activities from Barauni – Farrakka. APHA, AWWA and WPCF (1989) produced the standard methods for the examination of water and wastewater. Mahatre *et al.* (1980) studied the effect of industrial pollution on aquatic system of Kalu river. Badola and Singh (1981) studied the fish and fisheries of river Alaknanda. Mohanty (1981) studied the water quality of some water bodies of Bhubneshwar.

Horned (1982) investigated the water quality of Neuse river in terms of variability, pollution loads and long-term trends. Jeejabai and Rajendra (1982) observed the changing pattern of plankton composition as indicators of water pollution. Prasad and Yashpal (1982) emphasized diatoms as indicators of pollution. Daniel *et al.* (1982) analyzed water quality of French Broad river, based on water collected at Marshall. Patil and Patil (1983) investigated the water quality of river Ulhas with reference to Cadmium, Copper and Mercury. Chaphekar and Mahatre (1983) presented a report on the Impact of human settlement and development activities on Ganga river system. They interpreted that initial association of metals with the cells is governed by mere absorption. Raina *et al.* (1984) made pollution studies on river Jhelum. Somashekhar (1984) performed study on the river Cauvery. Chattopadhya *et al.* (1984) made pollution studies on the pollution status of river Ganga at Kanpur. Verma *et al.* (1984) observed the pollution and saprobic status of eastern Kalinadi. Trivedi and Goel (1984) presented a comprehensive work on chemical and biological methods for water pollution studies. Bhatt *et al.* (1985) studied the hydrobiology and phytoplankton population in river Kosi. Holander (1986) did microanalysis of scales of poecillid fishes and reported the teeth like structures called lepidonts. Sharma (1986) published a report on the effect of physico-chemical factors and benthic fauna of river Bhagirathi, Garhwal Himalayas. Aggarwal (1986) studied ecology of Chambal river at Kota, and Singh *et al.* (1983) presented geographical distributional list of icthyofauna of Garhwal Himalayas with some new records. Reddy and Venkateswarlu (1987) on the river Tungabhadra. Sivakumar *et al.* (1987) on the river Amravati provided valuable additional information on physico-chemical data. Dobriyal and Singh (1988) presented the ecological basis for icthyofaunal variations in two hill streams of Garhwal Himalaya. Nautiyal and Singh (1989) made observations on the river projects and the endangered Himalayan *Tor putitora* in the uplands of Garhwal region. Gautam *et al.* (1989) observed the seasonal variation of certain oxidation-reduction characteristics of the river Bhagirathi. Gautam (1990) studied the sediments of the river Bhagirathi, Garhwal Himalaya. Nautiyal (1990) explained the ecology of the Ganga river system in the upland of Garhwal Himalaya environment Khanna and Badola (1990) studied the icthyofauna of the river Ganga at foothills of Garhwal Himalayas. Borego *et al.* (1990) reported Cephalophage as an indicator of pollution. Moyle and Williams (1990) studied the biodiversity loss in temperate zone and reported the reasons for decline of native fish fauna in California and suggested methods for conservation.

Jhingran (1991) studied the fish and fisheries of India. Joshi *et al.* (1991) performed a study on the planktonic and benthic components of three selected tributaries of river Ganga between Devprayag and Rishikesh. Pathak (1991) studied some aspects of ecology of the rivers Sarju and Gomti of the Kumaun Himalaya. Joshi and Pathak (1991) performed a relative study of some physico-chemical parameters of sewage water at Uttarkashi. Singh (1991) published a book "Advances in limnology". Joshi *et al.* (1991) made an attempt to study diurnal variation in physico-chemical characteristics of Alaknanda river at Srinagar during winter season. Tombi (1992) studied the ecobiology of hill stream fishes of Manipur with special reference to culturable possibilities. Ramesh *et al.* (1992) evaluated the water quality in three rivers of Andhra Pradesh. Gautam (1992) suggested some considerations on stone removals from river basin and resultant effects. Mazhar and Kapoor (1992) made a limnological study on Doraniu river at Bareilly.

Saxena and Chauhan (1993) studied Physico-chemical aspects of pollution in river Yamuna. Khanna (1993) did an extensive study on ecology and pollution of river Ganga. Joshi *et al.* (1993) evaluated the physico-chemical characteristic of river Bhagirathi in the uplands of Himalayas. Khanna *et al.* (1993) studied about the physico-chemical and Biological parameter of river Ganga at Chandighat, Hardwar. Gautam *et al.* (1993) observed the physico-chemical changes and its impact on water quality of Alaknanda at Srinagar, Garhwal. Pandey *et al.* (1993) studied on the physico-chemical quality of water or river Kosi at Purnia.

Gautam *et al.* (1993) performed a study on physico-chemical characteristics of sewage and its impact on water quality of Alaknanda at Srinagar, Garhwal· Allan and Flecker (1993) made studies on the biodiversity conservation in running water. Joshi *et al.* (1993) studied the fish and fisheries of Gori Ganga. Murugesan *et al.* (1994) studied the water quality profile of the perennial river Tampraparni. Das (1994) suggested strategies for conservation of threatened fisheries. Joshi (1994) declared the status of fish conservation in river Ramganga. Grover *et al.* (1994) studied the icthyo-fauna of Doon Valley. Kumar (1995) studied on pollution in river Mayurakshi in south Bihar.

Gautam (1995) declared that continuous monitoring is an effective way to control water pollution. Das and Tiwari (1996) made observations on the seasonal variation in the productivity of some macro-invertebrates population of Bagulia Lake, Bihar. Abramovitz (1995) studies the fresh water failures. Shivsubramani and Mahadevan (1995) evaluated the river water quality of river Suruliyar in Tamil Nadu. Allum (1995) studied the structure and functioning of the running waters the various units of ecology and revealed that the decline in biodiversity is faster in aquatic environment particularly in streams and rivers. Joshi *et al.* (1995) analyzed Primary productivity in western Ganga canal at Haridwar. Chopra and Rehman (1995) made a study on self-purification of physico-chemical properties of Ganga canal water at Jwalapur, Haridwar. Shrivastava *et al.* (1996) evaluated the phytoplankton productivity and physico-chemical properties of Rapti river. Joshi (1996) studied the hydro biological profile of river Sutlej in its middle stretch in western Himalaya Joshi *et al.* (1996) observed the planktonic population in relation to certain physico-chemical factors of Ganga Canal at Jwalapur, Haridwar. Chopra and Arya (1996) did the physico-chemical and microbiological characterization of wastewater effluents of different sites of thermal power station, BHEL, Haridwar. Pandey *et al.* (1996) presented a study of diurnal variation in the physico-chemical parameters of two lentic water bodies of Darbhanga.

Khanna *et al.* (1997) observed the population of Green algae in relation to physico-chemical factors of the river Ganga at Haridwar. Kuhen *et al.* (1997) reported biochemical finger printing of Coliform bacteria population. He declared higher counts of bio-indicators at the confluence of polluted water in the river in down stream. Dutta *et al.* (1997) observed diel variations in physico-chemical characteristics of water and macro-benthic invertebrates in river Tawi, near Nagrota by pass, Jammu. Leidy and Moyle (1997) presented and over view conservation status of worlds fish fauna. Sharma and Pandey (1998) investigated the pollution status of Ramganga river at Moradabad. Bahumik and Saha (1998) told about the role of extension in arising mass awareness and public participation in fish conservation movement. Khanna *et al.* (1998) performed an extensive study on the fish fauna of river Ganga at Rishikesh. They reported that during their course of study various species of fishes as *Tor tor*, *Tor putitora* and *Raimas bola* were found endangered and the main reason was declared to be the polluted habitats, land slides and soil erosions besides this, the destructive methods of fishing in the river Ganga is also the prime factor of depleting fish fauna. Kaur *et al.* (1999) made an attempt to study the aquatic diversity of Kalanji Lake, Punjab. Badola and

Badola (1999) gave their observations on destructive fishing methods and recommendations to protect fish fauna in Garhwal Himalayas. Kumar and Bohra (1999) observed Gastropods as indicators of pollution status of some wetlands in Santal pargana, Bihar. Khanna *et al.* (1999) studied the fish and their ecology in river Suswa at Raiwala. Kumar and Mukherjee (2000) observed the changes in mammalian and fish fauna of Doon Valley and surrounding area. Karuppasamy (2000) studied the short and long term effect of Phenyl mercuric acetate on the protein metabolism in *Channa punctatus* (Bloch). Dua and Johal (2000) used scales of *Channa punctatus* (Bloch) as an indicator for pesticide pollution and declared significant impairment upon exposure to this organochlorine pesticide. They said that on exposure to endosulfan there was disruption in lepidonts from the anterior side of the scales, thus result in loosening of hold of scales from the body of fishes. Khanna *et al.* (2000) studied the impact of abiotic factors on the planktonic population of a pond in Haridwar. Gautam *et al.* (2000) monitored the quality deterioration in aquatic environment of the Ganges: A growing problem for future. Gautam *et al.* (2000) investigated the diurnal variations in the physico-chemical characteristics of Ganga water at Triveni ghat, Rishikesh during winter season. Sharma *et al.* (2000) studied the characteristics and pollution effects of pulp and paper mill waste on Tapti river, District Surat. Gautam *et al.* (2001) attempted the study of quality deterioration in aquatic environment of the Ganges. Shanthi *et al.* (2001) made a hydro biological study of Singanallur lake at Coimbatore, India. Sharma *et al.* (2002) studied the benthic community of Sirpur Lake, Indore. Khanna *et al.* (2002) studied the impact of paper mill effluent on some water parameters of Hindon river, at Saharanpur. Kaur and Joshi (2002) observed the solid waste generation during normal days in Haridwar city. Mishra and Joshi (2002) observed the generation of solid waste during festive occasions in Haridwar city. Abraham *et al.* (2002) made efforts to study the distribution of hydrographic parameters and heavy metal ions in river Periyar.

Singh *et al.* (2002) observed the distribution and seasonal abundance of benthic macro invertebrates in the Bisar sarovar of Gaya, Bihar. Pathani *et al.* (2002) described some physico-chemical parameters and primary productivity of river west Ramganga. Bahuguna and Badoni (2002) made a study on qualitative and quantitative productivity of fish food from three important tributaries of river Alaknanda, Garhwal. Dutta *et al.* (2002) presented an overview of the fish diversity of Ranchi district. Fokmare and Musaddique (2002) investigated the physico-chemical characteristics of Kapashi lake and Purna river waters in Alok district of Maharashtra. Kumar and Sharma (2002) studied the water quality of Krishna river (Physico-chemical parameters). Sukumaran (2002) made observation on Coliform bacteria an index of aquatic pollution. Basu and Mitra (2002) studied the pollution of river Damodar at Burdwan with special reference to metals, COD and pH. Kumar (2002) observed the effects of ammonia to fresh water murrel, *Channa punctatus* (Bloch). Ahmed and Begum (2002) monitored the bacteriological quality of water in hill stream of Khanpara during different seasons.

Khanna and Bhutiani (2003) observed the limnological characteristics of river Ganga at Haridwar. Khanna and Bhutiani (2003) studied the limnological status of Satikund pond at Haridwar. Malik *et al.* (2003) observed the impact of distillery effluent on the water quality and phytoplankton in river Saung Prasad (2003) made observation on physico-chemical and bacteriological quality of ground waters at Tadepalli Mandal of Guntur district, Andhra Pradesh. Sivakumar *et al.* (2003) studied the water quality of the river Ambarampalayam, Coimbatore District, Tamilnadu. Monichan and Sharma (2003) studied the distribution status of fish fauna in Etawah district, U.P.

Jena *et al.* (2003) performed studies on water quality parameters around Sagar Island, Sundarbans. Wanganeo *et al.* (2003) revealed that lotic system under temperate climatic condition support less fish diversity in comparison to tropical climatic conditions. Sivakumar *et al.* (2003) presented a study on some biological aspects of fresh water fish. Khanna and Bhutiani (2004) studied ecology of some fishes of river Ganga at Gohri Ghat, Garhwal in relation to abiotic factors. Esmaeili *et al.* (2001) performed SEM studied on body scale of a tooth carp, *Aphanius ginaonis* (Holly) from the hot spring of Geno, South Iran· Joshi (2005) made study on the ecology and contribution of smaller mountain tributaries to the productivity of major river Bhagirathi-Ganga- in Himalayan region. Johal *et al.* (2005) declared that streams are a great threat because of anthropogenic activities like excavation of bed material, construction of dams, channelization, siltation, agricultural activities and majorly pollution. Johal *et al.* (2005) made a study on the rivers of major landscape features, representing visible portion of the aquatic system. Jindal and Singh (2005) also studied the ecology of benthos of some fresh water systems of Punjab.

CHAPTER-3

MATERIALS AND METHODS

For present study of River Ganga the water samples were collected seasonally from different sampling stations (A, B, C, D and E) during October 2003 to September 2005 in morning hours (from 7.00 A.M. to 10.00 A.M.). The samples were taken in borosil glass bottles of 300ml, plastic cans of 1 lt. from each location.

The samples for different parameters were analysed with the help of the procedure described by Welch (1948), APHA (1989), Trivedi and Goel (1984) and Khanna and Bhutiani (1995).

SAMPLING METHODS

A. CLEANING AND STERILIZATION OF GLASSWARES

For cleaning the glasswares washing soda or chromic acid was used.

1. Sample Bottle :

300 ml capacity BOD bottles made of borosil were used. They were washed with washing soda or chromic acid and rinsed with tap water followed by distilled water. The neck and stopper were wrapped by butter paper with the help of rubber band. Sterilization of the sample bottle was done in autoclave at 15 lbs/inch^2 pressure and 121°C for 20 minutes.

2. Pipettes :

Different volume size pipettes were washed and fitted with cotton plug at the upper end. These were wrapped in butter paper and sterilization was done in autoclave at 15 lbs/inch^2 pressure and 121°C for 20 minutes.

3. Petridishes :

Borosil 90mm diameter petriplates were washed and then sterilization was done in an oven at 180°C for 2 hours.

4. Test tubes :

Borosil test tubes were washed and then plugged with non–absorbent cotton wool. These were arranged in test tube racks and sterilization was done in autoclave at 15 lbs /inch^2 pressure and 121° C for 20 minutes.

B. SAMPLING

For physico-chemical and microbiological parameters, Ganga river water samples were collected from five sites in plastic cans. Sampling for dissolved oxygen (DO) and bacteriological analysis was done separately.

1. Sampling for Dissolved Oxygen (DO) :

The samples were collected in clean and sterilized 300 ml capacity BOD bottles. The bottles were filled completely with Ganga water and stoppered was placed inside the water only. Immediately DO was fixed by adding 2 ml of each alkaline KI and manganous sulphate ($MnSO_4$) at the sampling site.

2. Sampling for bacteriological analysis :

Cleaned and sterilized sampling tubes were used. It was dipped into the water with its neck downward upto 6 inches below the water surface and then its mouth was opened under water against the flow. The tube was not filled completely and about 10% of the volume of the tube was kept empty. It was closed immediately taking out of the water. To avoid contamination of sample, the neck and the stopper were wrapped by butter paper with the help of rubber band.

C. WATER ANALYSIS METHODS

The Physico-chemical and microbiological analysis of water samples were made by the following methods out lined in Welch (1948), APHA (1989), Mathur (1982), Trivedi and Goel (1984) and Khanna (1993). Samples were analysed for the following parameters.

(I) PHYSICO-CHEMICAL PARAMETERS :

1. Temperature
2. Conductivity
3 Turbidity
4 Velocity
5. Total Solids (TS)
6. Total Dissolved Solids (TDS)
7. Total Suspended Solids (TSS)
8. pH
9. Dissolved Oxygen (DO)
10. Free Carbon dioxide (CO_2)
11. Alkalinity
12. Biochemical Oxygen Demand (BOD)
13. Chemical Oxygen Demand (COD)
14. Hardness
15. Acidity
16. Chloride (Cl^-)
17. Calcium
18. Total Nitrogen
19. Magnesium

20. Heavy Metals
    (a) Copper
    (b) Iron

(B) BIOLOGICAL PARAMETERS

21. Plankton
22. Micro Benthicfauna
23. Fish

(C) MICRO-BIOLOGICAL PARAMETERS:

24. Bacterial Identification and Isolation
25. Standard Plate Count (SPC),
26. Most probable number (MPN)

Note: Temperature, pH, CO_2, DO, BOD, SPC and MPN were analyzed immediately after sampling.

METHODOLOGY

1. Temperature

This parameter is basically important for its effects on the chemistry and biological reactions of the organisms in the water.

The thermometer is the best suitable device for measuring the temperature. The bulb of thermometer dipped into the water and the reading was noted in °C (Celsius).

2. Turbidity

The substances not present in the form of true solution cause turbidity in water. True solutions have a particle size of less than 10^{-9} m. Any substance having more than this size will produce a turbidity. Turbidity of water is actually the expression of optical property (Tyndall effect) in which the light is scattered by the particles present in the water.

Clay, silt, organic matter, phytoplankton and other microscopic organisms cause turbidity in natural waters. Turbidity makes the water unfit for domestic purpose, food and beverage industries, and many other industrial uses.

The turbidity of the water was estimated by Jackson's candle turbidimeter. The water samples were collected from the river in plastic cans, collected sample were shaked well and transferred into the glass tube of Jackson's candle turbidimeter until the flame of the candle disappears and noted the light path in centimeter (measured from inside the bottom of the glass tube), and the reading recorded in JTU.

3. Conductivity

Electrical conductance is the ability of a substance to conduct the electric current. In water, it is the property caused by the presence of various Ionic species. The ELICO conductivity meter determined conductivity of the sample.

4. Velocity

The speed of water current was measured by the surface float method. The rate of flow of the uppermost stratum of the water in the river was measured by throwing a light piece of wood in different segments of the river. Then the time take by the wood piece to cover a known distance was calculated by given formula:-

$$V = \frac{d}{t}$$

where V = velocity (m/sec.)

d = distance covered by the float (metre)

t = time take by the float (sec.)

5. Total Solids (TS)

Total solids or total residue is the term applied to material left in a beaker after evaporating a well mixed sample and subsequently evaporating and dried it in a oven.

Procedure :

A 100 ml washed and dried beaker was taken and weighed. 100 ml of unfiltered sample was putted in the beaker and was evaporated on water bath then dried it in an oven at 105 ± ± 1°C. The final weight of the beaker was taken after drying of sample.

Calculation :

$$\text{Total solids, mg/L} = \frac{A-B}{V} \times 10^6$$

where,

A = Final weight of beaker in g

B = Initial weight of beaker in g

V = Volume of the sample taken in ml.

6. Total Dissolved Solids (TDS)

Total dissolved solids or filterable residue are those solids, which left after evaporation of the filtered sample.

Procedure :

A 100 ml washed and dried beaker was taken and weighed immediately before use. The filter paper Whattman No. 42 was placed in funnel beaker and the 100 ml of well-mixed sample was filtered through filter paper. The filtrate was collected in the 100 ml weighed beaker. After complete filtration, the sample was evaporated and subsequently dried in an oven at 105 ± ± 1°C for 4 to 6 hrs. Finally the beaker was cooled and weighed.

Calculation :

$$\text{Total dissolved solids, mg/L} = \frac{A-B}{V} \times 10^6$$

where, A = Final weight of beaker in g.

B = Initial weight of beaker in g.

V = Volume of the sample taken in ml.

7. Total Suspended Solids (TSS)

Total suspended solids or non-filterable residue are the retained material on Whattman No. 42 filter paper after filteration of well mixed sample.

Procedure :

TSS was determined by taking difference between the total solids and total dissolved solids.

Calculation :

TSS, mg/L = TS – TDS

8. Potential Hydrogenii (pH)

pH is the hydrogen ion activity or concentration in the given water sample. pH equals to negativity $\log_{10}$ of hydrogen ion concentration.

$$pH = -\log_{10} (H^+)$$

pH can be determined by the pH strips, but the accurate method is the electrometric method in which potential in between the two electrodes (indicator and calomel) is measured directly. pH of the river water sample was measured with the help of electrodes of pH meter (ELICO). The apparatus has a reference and an indicator glass electrode.

Procedure :

The apparatus was calibrated with the help of buffer solutions having known pH 4.01 and 9.18 called as pthalate and borate buffers respectively. Now both of the electrodes were rinsed with distilled water and then dipped in the test solution taken in a 100ml glass beaker. As the electrodes were dipped in the sample, the pH meter gives the value of the test solution.

Reagents :

(a) Potassium hydrogen pthalate buffer :

10.2 g. of potassium hydrogen pthalate was dissolved in distilled water to prepare 1000 ml of buffer.

(b) Borax buffer :

3.81 gm of $Na_2B_4O_7.10H_2O$ was dissolved in distilled water to prepare 1000 ml of buffer.

9. Dissolved Oxygen (DO)

DO is the measure of oxygen concentration present in a given water sample. The concentration of oxygen reflects whether the process undergoing is aerobic or anaerobic. There are two ways for the detection of the DO.

A. Winkler's Idometric method

B. Azide modification of winkler's method

Although there is minute change between the two methods, but we followed the second.

Azide Modification of Winkler's Iodometric Method :

Principle :

The manganous sulphate reacts with the alkali (KOH or NaOH) to form a white precipitate of manganous hydroxide, which in the presence of oxygen gets oxidized to brown colour compound. In the strong acid medium, manganic ions are reduced by iodide ions, which get converted to iodine, equivalent to the original concentration of the oxygen in the sample. The iodine can be titrated against this sulphate using starch as indicator.

Certain oxidizing and reducing materials are effectively interfere with the determination of oxygen by converting iodide on ions to iodine or vice-versa. Azide modification removes such substance specially nitrite, which is distorted by sodium azide (NaN_3).

Reagents

(a) Sodium thiosulphate, 0.25N :

24.82g of $Na_2S_2O_3.5H_2O$ was dissolved in boiled distilled water and madeup the volume to 1 ltr. 0.4g of borax or a pallet of NaOH was added as stabilizer. This was 0.1N stock solution. It was diluted 4 times with boiled distilled water to prepare 0.25 N solutions. It was kept in a brown glass stoppered bottle.

(b) Alkali iodide azide solution :

(i) 500 gm of NaOH or 700 gm of KOH and 150 gm of KI were dissolved to make 1 lt. of solution.

(ii) 10gm of NaN_3 was dissolved in distilled water.

Two solution (i) and (ii) were mixed.

(c) Manganous sulphate solution :

100 gm of $MnSO_4.4H_2O$ was dissolved in 200 ml of boiled distilled water and filtered.

(d) Starch solution :

1 gm of starch was dissolved in 100 ml of warm distilled water and a few drops of formaldehyde solution was added.

(e) Sulphuric acid :

H_2SO_4, conc. (sp. gr. 1.84)

Procedure :

The sample was filled in a glass-stoppered bottle (BOD bottle) of 300 ml capacity, avoiding any kind of bubbling and trapping of air bubble after placing the stopper. 2 ml of each $MnSO_4$ and alkaline KI were poured into the bottom of sample by the pipette. A precipitate was appeared. The stopper was placed and shook the contents well by inverting the bottle repeatedly. The bottle was kept for some time. This is the stage, where the DO was fixed. Now 2 ml of conc. H_2SO_4 was added and shook to dissolve the precipitate and titrated the whole content against hypo solution using starch as an indicator, till the blue colour changed to colourless. Now the DO was calculated.

Calculation :

$$\text{Dissolved Oxygen, mg/L} = \frac{(\text{ml} \times \text{N}) \text{ of titrant} \times 8 \times 1000}{V_1 - V}$$

where, V_1 = Volume of bottle after placing the stopper

V = Volume of $MnSO_4$ and KI added

10. Free Carbon Dioxide (CO_2)

Principle :

Free CO_2 can be determined by titrating the sample using a strong alkali (Such as carbonate free NaOH) to pH 8.3. At this pH all the free CO_2 is converted into bicarbonates.

Reagents :

(a) Sodium hydroxide, 0.05 N :

1.0N NaOH was prepared by dissolving 40 gm of NaOH in CO_2 free distilled water (boiled) to make 1.0 litre of solution. 50 ml of 1.0 N NaOH was diluted to make 1 litre. Standardization of the solution was done with H_2SO_4.

(b) Phenolphthalien indicator :

0.5 gm of phenolphthalein was dissolved in 50 ml of 95% ethanol and 50 ml of distilled water was added. Drop by drop 0.05N CO_2 free NaOH solution was added, until the solution was turned faintly pink.

Procedure :

100 ml of sample was taken in a conical flask and a few drops of phenolphthalein indicator was added. If the colour turns pink, free CO_2 declared to absent. If the sample remained colourless titrate it against 0.05N NaOH. At the end point a pink colour was appeared.

Calculation :

$$\text{Carbon dioxide } (CO_2)\text{, mg/L} = \frac{(\text{ml x N}) \text{ of NaOH x 1000 x 44}}{\text{ml of sample}}$$

11. Alkalinity

It is the measure of the capacity of the water to neutralize a strong acid.

Principle :

Total alkalinity can be estimated by titrating the sample with a strong acid (HCl or H_2SO_4), first to pH 8.3 using phenolphthalein as an indicator and then further to pH between 4.2 and 5.4 with methyl orange. In the first case the value is called as phenolphthalein alkalinity (PA) and in second case, the value is called as total alkalinity (TA). Values of carbonates, bicarbonates and hydroxyl ions can be computed from these two types of alkalinities.

Reagents :

(a) Hydrochloric acid, 0.1 N :

12 N concentration HCl (Sp. Gr. 1.18) was diluted to 12 times (8.34→100ml) to prepare 1.0 N HCl. It was further diluted ten times to make 0.1 N HCl (100→1000 ml).

(b) Phenolphthalein indicator :

0.5 g of Phenolphthalein was dissolved in 50 ml of 95% ethanol and 50 ml of distilled water was added. Drop by drop 0.05 N CO_2 free NaOH solution was added, until the solution was turned faintly pink.

(c) Methyl Orange Indicator, 0.05 % :

0.05 g of methyl orange was dissolved in 100 ml of distilled water.

Procedure :

100 ml of sample was taken in a conical flask and 2 drops of phenolphthalein was added (i) solution remained colourless, so phenolphthalein alkalinity was absent (PA = 0) and total alkalinity was determined. (ii) If the colour changes to pink, it is titrated with 0.1N HCl until the colour disappeared at the end point. This is phenolphthalein alkalinity (PA).

Now, 2 drops of methyl orange was added to the same sample and titration was continued with 0.1N HCl, until yellow colour changed to pink. This was total alkalinity (TA).

Calculation :

$$\text{PA as CaCO}_3\text{, mg/L} = \frac{(\text{A x normality of HCl}) \times 1000 \times 50}{\text{Volume of sample (ml)}}$$

$$\text{TA as } CaCO_3\text{, mg/L} = \frac{(\text{B x normality of HCl}) \times 1000 \times 50}{\text{Volume of sample (ml)}}$$

where, A = ml of HCl used with only phenolphthalein

B = ml of total HCl used with only phenolphthalein and methyl orange

12. Biochemical Oxygen Demand (BOD)

BOD is the measure of biodegradable organic matter present in a water sample and can be defined as the amount of oxygen required by the microbes in stabilizing the biologically degradable organic matter under aerobic conditions. Thus BOD value can be used as a measure of waste strength.

Principle :

The principle of the method involves, measuring the difference of the oxygen concentration between the sample and after incubating it for 5 days at 20°C.

Apparatus and Reagents :

(a) BOD bottles

(b) BOD incubator

(c) Phosphate buffer

(d) Magnesium sulphate

(e) Calcium chloride

(f) Ferric chloride

(g) Sodium thiosulphate, 0.25 N

(h) Alkali iodide azide solution

(i) Manganous sulphate solution

(j) Starch solution

(k) Sulphuric acid

Dilution water was prepared by bubbling compressed air in distilled water for about 30 minutes. Now 1 ml. each of $MgSO_4$, $FeCl_3$, PO_4 buffer and $CaCl_2$ were added into one litre of water and mixed thoroughly. The sample was neutralized around pH

7.0. As the DO in the sample is likely to be exhausted, so a suitable dilution of the sample was prepared according to the expected BOD range. This dilution sample water filled the two sets of BOD bottles. One set of the bottles was kept in BOD incubator at 20°C for 5 days and the DO constant was determined in another set immediately. After the compilation of 5 days incubation, the DO of first set was determined BOD was calculated by the given formula.

Calculation :

BOD, mg/L = $(D_0 - D_5)$ × dilution factor

Where, D_0 = Initial DO in the sample

D_5 = DO after 5 days

13. Chemical Oxygen Demand (COD)

COD is widely used to characterize the organic strength of wastewaters and pollution of natural waters. The test measures the amount of oxygen required for chemical oxidation of organic matter in the sample to carbon dioxide and water.

Principle :

The sample is refluxed with $K_2Cr_2O_7$ and H_2SO_4 in presence of $AgSO_4$ (catalyst) and $HgSO_4$ (to neutralize the effect of chlorides). The excess of $K_2Cr_2O_7$ is titrated against ferrous ammonium sulphate using ferroin as an indicator. The amount of $K_2Cr_2O_7$ used is proportional to the oxidizable matter present in the sample.

Reagents :

(a) Potassium dichromate solution, 0.025 N :

12.259 g of dried A.R. grade $K_2Cr_2O_7$ was dissolved in distilled water to make 1 litre of solution. This is 0.25 N $K_2Cr_2O_7$ solution. It was diluted 10 times to make 0.025 N $K_2Cr_2O_7$ solution.

(b) Ferrous ammonium sulphate, 0.01N :

39.2 g of $FeSO_4(NH_4)_2SO_4.6H_2O$ was dissolved in water, adding 20 ml conc. H_2SO_4 to make 1 litre of 0.1N solution.

It was diluted 10 times (100→1000 ml) to make 0.01 N $FeSO_4(NH_4)_2SO_4.6H_2O$ solution.

(c) Ferroin indicator :

1,10-phenonthroline and 0.695 g of ferrous sulphate ($FeSO_4.7H_2O$) were dissolved in distilled water to make 100 ml of solution.

(d) Sulphuric acid :

H_2SO_4 conc. (Sp. Gr. 1.84)

(e) Mercuric sulphate :

$HgSO_4$ solid

(f) Silver sulphate :

Ag_2SO_4 solid

Procedure :

20 ml of sample was taken in COD flask. 10 ml of 0.025 N $K_2Cr_2O_7$ and a pinch of Ag_2SO_4 and $HgSO_4$ were added. (In case of the COD is expected below 50 mg/l, 10ml of 0.025N $K_2Cr_2O_7$ should be added). Now 30 ml of conc. H_2SO_4 was added and reflux condenser was connected to the COD flask. It was refluxed for 2 hours on water bath. The flask was removed, cooled and distilled water was added to make the final volume to about 140 ml, 2-3 drops of ferroin indicator was added and titrated with 0.1 N $FeSO_4(NH_4)_2SO_4$.

A blank was also run with distilled water, using same quality of chemicals.

Calculation :

$$\text{COD, mg/L} = \frac{(B - A) \times \text{N of FeSO}_4\ (\text{NH}_4)_2\ \text{SO}_4 \times 1000 \times 8}{\text{Volume of sample (ml)}}$$

where, A = Volume of titrant with sample (in ml)

B = Volume of titrant with blank (in ml)

14. Total Hardness

Total hardness was determined by using EDTA titrametric method. A 100 ml sample of river water was titrated by 0.01N EDTA solution after the addition of indicator (very small amount of EBT± 1 ml of ammonium buffer). End point was recorded at the time of first colour change (pinkish to blue) and total hardness was then calculated by using the formula.

Calculation :

$$\text{Hardness (mg/L)} = \frac{\text{ml of EDTA used} \times 1000}{\text{ml of sample}}$$

15. Acidity

Acidity of water is its capacity to neutralize a strong base and is moisty due to the presence of strong minerals acids, weak acids, (Carbonic, acetic acids) and the salts of strong acids and weak bases. These salts on hydrolysis produce a strong acid and metal hydroxide, which are sparingly soluble thus producing the acidity.

Take 100ml of colourless sample in a conical flask and add 2-3 drops of methyl orange indicator, if solutions turn yellow methyl orange acidity is absent. If solution turns pink titrate it against 0.05N NaOH until colour changes to yellow. Now add few drops of phenolphthalein indicator and titrate against 0.05N NaOH at end point solution turns pink.

$$\text{Total acidity} = \frac{\text{A} + \text{B of NaOH} \times 1000 \times 50}{\text{ml of sample}}$$

where, A = Volume of NaOH used with methyl orange in titratig the sample to pH 3.7

B = Volume of NaOH used with phenolphthalein in titration the sample from pH 3.7 to pH 8.3.

16. Chloride (Cl^-)

The most important source of chloride in the water is industries and discharge of domestic sewage. The chloride concentration serves as an indicator of pollution by sewage.

Principle :

Silver nitrate reacts with chloride to form very slightly soluble white precipitate of silver chloride (AgCl). At the end point where all the chloride gets precipitated, free silver ions react with chromate to form silver chromate of reddish brown colour.

Reagents :

(a) Silver nitrate, 0.02 N :

3.4 gm of dried $AgNO_3$ (A.R.) were dissolved in distilled water to make 1 litre of solution and kept in a dark bottle.

(b) Potassium chromate 5% :

5 gm of K_2CrO_4 was dissolved in distilled water.

Procedure :

50 ml of sample was taken in an Erlenmeyer flask and added 2 ml of K_2CrO_4 solution. The contents were titrated against 0.2N $AgNO_3$ until a red tinge appeared.

Calculation :

$$\text{Chloride, mg/L} = \frac{(\text{ml} \times \text{N}) \text{ of } AgNO_3 \times 1000 \times 35.5}{\text{Volume of sample (ml)}}$$

17. Ca hardness

100ml of sample was taken in a flask, followed by the addition of mure oxide indicator and 11ml of 1N NaOH. After it the sample was titrated by standard EDTA solution till the color changed (pink colour change to purple) this is end point. This is calculated by understated equation:

$$\text{Ca hardness (mg/l)} = \frac{(\text{ml of EDTA used} \times 1000}{\text{ml of sample titrated}}$$

18. Total Nitrate

It was measured by phenol disulphonic acid method. A 50 ml. sample was taken in a flask followed by the addition of 50ml of Ag_2SO_4 solution to remove chlorides. Then, it was kept in an oven for 15 minutes, filtered and evaporated to dryness. Residue thus obtained, was dissolved in 2ml of disulphonic acid, and diluted to yellow color. The color intensity was measured in spectrophotometer at 410 nm and the amount was calculated from the standard curve prepared at the time of analysis.

19. Mg hardness

Magnesium was estimated by subtracting the calcium hardness from total hardness.

It was calculated by using formula:-

Mg hardness (mg/ l) = Total hardness – Ca hardness.

20. Heavy Metal: Copper and Iron

By Atomic Absorption Spectrophotometry (AAS) (ECIL).

Principle of AAS :

AAS consisting of light source emitting the line spectrum of element (hallow cathode lamp or electrode less discharge lamp), a device for evaporizing the sample (using a flame, a means of isolating an absorption line, monochromator or filter and adjustable slit) and a photoelectric detector with its associated electronic amplifying and measuring equipment.

Method for using AAS :

Requirement for determining by atomic absorption spectrophotometry (AAS) vary with and/or concentration to be determined. In atomic absorption spectrophotometery a sample is absorbed into a flame and atomized and a light beam is directed through the flame, into a monochromometer and on to a detector that measure the amount of light absorbed by the automized element in the flame. Each metal has its own characteristic absorption wavelength. A source lamp composed of that element is used. The amount of energy of the characteristic wavelength absorbed in the flame is proportional to the concentration of element in the sample. For the detection of sodium and potassium the sample was directly aspirated into air acetylene flame and the concentration recorded.

The instrument was calibrated, for each metal to be detected, with known standard. After calibration and programming, the prepared samples were aspirated to the flame. Zero was set with double distilled water before each determination. Results are displayed in mmgm/l from which the actual concentration in water samples was calculated. It was the total content of water sample, which was expressed in mg/l (ppm).

(B) BIOLOGICAL PARAMETERS:

21. Plankton

Collection of plankton :

The plankton of river were collected by fine plankton net. The known quantity of water was pass through filter, and concentrated in the desired quantity, 1 ml of the concentrate was taken and placed in a Sedgwick Rafter counting cell. Applying the following formula did counting of the organism.

$$\text{Plankton/l} = \frac{(a \times 1000)\,C}{1}$$

where, a =average no. of phytoplankton in one small counting chamber of Sedgwick Rafter counting cell

c = ml. of plankton concentrate

1 = Volume of original water filtered in litre.

22. Micro Benthicfauna

Zoo benthoses are larger sized organisms present in the water body. The macro invertebrates found in fresh water are annelids, gastropods and insects. These are of two types either Epifaunic (Benthic forms that move about freely or are attached) or Infaunic (Benthic forms that dig and live in burrows in soil). Collection of benthos was done with the help of dragnet, insect net and forceps were used. The collected materials were preserved in 5% formaline and identified with the help of Ward and Whipple (1959), Needham and Needham (1972) and APHA (1989).

23. Fish

Fishes were collected from river by jal, fatiyala (triangular net) and fine clothes. All the series specimens were tagged, numbered, preserved in 5% formaline and identified with the help of Badola (1979) and Jhingran (1975).

(C) MICROBIOLOGICAL PARAMETERS

24. Standard Plate Count (SPC)

On a solid medium, counting of organism depends upon the fact that living cells will produce sufficient progeny to form a colony visible to naked eyes. Since bacteria occuring in water are single cell, pairs, groups or even dense clumps, not every individual living cell will develop into a separate colony on incubation. Therefore a number of colonies appearing on plate does not necessarily represent the total number of organisms present in the test volume. The results are expressed as the cell forming units (CFU) per ml.

Apparatus and Materials

(a) Incubator
(b) Petridishes
(c) Conical flasks (250 ml)
(d) Biological pipettes
(e) Colony counter
(f) Water bath
(g) Culture tubes
(h) Nutrient Agar Medium (NAM)

Nutrient Agar Medium (NAM)

Beef extract	-	3.0 g
Peptone	-	5.0 g
Agar agar	-	15.0 g
Distilled water	-	1.0 L
pH	-	6.8 ± ± 0.2
(after sterilization)		

Preparation of serial dilution :

5 or 6 culture tubes each containing 9ml. distilled water, were sterilized after plugging. The sample was vigorously shaken and 1 ml was transferred to the first culture tube. Now 1 ml water from first tube was again transferred to 2nd test tube and so on. Separate pipettes were used for each transfer. The dilutions were 1:10, 1:100,1:1000, 1:10000 and so on.

Procedure :

3 pairs of borosil petridishes were sterilized in an oven. Now 1 ml of liquid from last 3 dilution was transferred to the petridishes by opening them gently. 15-20 ml medium was poured into these plates. The temperature of NAM should be about 45°C. Both water sample and medium, were mixed by gently rotating the plate. When medium became solidify the plates were inverted and incubated at 37± ± 2°C temperatures for 48 hours. The colonies were counted after 48 hours by colony counter and Colony Forming Unit (CFU)/ml was calculated.

Calculation :

$$\text{CFU/ml} = \frac{\text{Colonies Counted}}{\text{Dilution Factor}}$$

25. Most Probable Number (MPN)

The bacteriological quality of water is based on testing for bacterial species of known excretal origin, particularly organisms of the coliform group. The coliform group of bacteria included all the aerobic and facultatively anaerobic, Gram negative, non-sporulating bacilli, that produce acid and gas from the fermentation of lactose.

Coliform bacteria as typified by *E. coli* and faecal *Streptococci* residing in the intestinal tract of humans are excreted in faeces of humans and other warm-blooded animals averaging about 50 million coliforms per gram.

The multiple tube fermentation technique can be used to enumerate positive presumptive and confirmed coliform tests. Results of the test are expressed as the Most Probable Number (MPN), since the count is based on statistical analysis of sets of tube in a series of serial dilution. MPN is by definition is related to a sample volume of 100 ml.

Apparatus and Material

(a) Fermentation tubes (50 ml)

(b) Durham's tubes

(c) Biological pipettes

(d) Inoculating loop

(e) Incubator

(f) Conical flasks

(g) Autoclave

(h) Spirit lamp

(i) Water bath

ii. PRESUMPTIVE TEST :

MacConkey's Broth is required for this test. The composition of broth is as given below:

Bile salt or Sodium taurocholate	-	5.0 g
Peptone	-	20.0 g
Lactose	-	10.00 g
NaCl	-	5.0 g
1% Bromo cresol purple	-	1.0 ml
Distilled water	-	1.0 litre
Final pH (after sterilization)	-	7.1 ± ± 0.2

Procedure :

Several dilutions of the sample for river water were prepared. Dilutions above than 1:1000 are required. 5 fermentation tubes for each dilution, were filled with 9 ml of broth and one Durham's vial was put in each tube in inverted condition. Now all the tubes were plugged with cotton and sterilized in autoclave. All the dilutions were shaked vigorously before moving the sample to fermentation tubes. Separate sterilized pipettes were used for different dilutions as well as samples. Fermentation tubes were gently shaked. Now these tubes were incubated at 35-37°C temperature for 48 hours. All the tubes were examined after 48 hours for gas production.

The tube showing gas production in Durham's tube, were recorded as positive (± ve). Even a tiny bubble should be recorded as positive. The tubes showing positive results were now subjected to the confirmatory test. Negative tubes were incubated for further 24 hours to confirm the test.

i. Confirmatory test :

For this test Brilliant Green Lactose Bile (BGLB) Broth is required. This inhibits the growth of lactose fermenters other than coliform; thus, gas formation in BGLB broth constitutes a confirmed test i.e. coliform present. The composition of BGLB broth is given below.

Peptone	-	10.0g
Lactose	-	10.0g
Oxgall	-	20.0g
Brilliant green	-	0.0133g
Distilled water	-	1.0l
pH	-	7.2 ± 0.2
(after sterilization)		

Procedure :

Fermentation tubes with 10ml BGLB broth and inverted Durham's tube were prepared. The number of tubes should be equal to all ± ve tube in the presumptive test. Positive tubes were shook gently. One loopful of liquid was transferred to the tubes having BGLB with the help of inoculation loop. These tubes were incubated at 35–37°C for 48 hours. The tubes were recorded as positive, which showed the gas production. The number of positive tubes was put in MPN table and then in the formula to calculate the MPN/100ml.

Calculation :

$$\text{MPN/100 ml} = \frac{\text{MPN table value} \times 10}{\text{Starting value}}$$

26. Isolation and Identification of bacterial culture

Isolation of Bacteria :

Two methods were used to isolate bacteria and are

(i) Pour Plate Method

(ii) Streak Plate Method

(i) Pour plate method :

In this method 5 or 6 culture tubes each containing 9 ml distilled water, were sterilized after plugging for serial dilution. The sample was vigorously shaken and 1 ml was transferred to the first culture tube. Now 1 ml water from first tube was again transferred to 2nd test tube and so on. Separate pipettes were used for each transfer. The dilutions were 1:10, 1:100, 1:1000, and so on.

3 pairs of borosil petridishes were sterilized in an oven. Now 1 ml of liquid from last 3 dilutions was transferred to the petridishes by opening them gently 15–20 ml medium was poured into these plates. The temperature of NAM should be about 45°C. Both water sample and medium were mixed by gently rotating the plate. When medium became solidify the plates were inverted and incubated at 37 ± 2°C temperature for 48 hours.

(ii) Streak method:

From these poured plates, individual colony was transfered with the help of sterilized loop (by incineration) on another media plate and streaked. These plate were incubated for further 48 hours at 37 ± 2° C, for identification of pure colony.

Identification of Bacteria :

(i) Incubation Temperature :

Each isolated colony was incubated at below 20°C (Psycrophileic), 20 – 45°C (Mesophileic), and 45°C and above (Thermophileic).

(ii) Gram Staining :

Individual isolated colonies were stained by Gram stain and differentiated as Gram + ve and Gram –ve.

Principle :

Gram staining is a method of differential staining, which employs two stains to distinguish usually morphologically similar but different species of bacteria. The technique involves staining of bacteria by crystal violet (Primary stain) and safranin or basic fuchsin (Counter stain). The bacteria retaining the primary stain are called gram ± ve, while the bacteria yielding out the primary stain by alcohol washing and retaining the counter stain are called gram –ve.

Reagents :

A. Crystal Violet

   (i) Dissolve 2g of crystal violet in 20ml of 95% ethyl alcohol.

   (ii) Dissolve 0.8g of ammonium oxalate in 80ml of distilled water.

   Mix the two solutions (a and b).

B. Gram's Iodine

   Dissolve 1g of iodine and 2 g of KI in 300ml of distilled water.

C. Ethyl alcohol, 95% or Acetone alcohol (1:1)

D. Safranin, 0.5%

Procedure :

(1) Prepare a suspension of bacterial colony in sterile 0.85% NaCl (Saline) to disperse the bacterial cells.

(2) Take a slide and mark an area on it by a glass marking pencil.

(3) Prepare a smear of the bacterial suspension in the area marked on the glass slide. Air-dry and fix the smear by very slight heating.

(4) Pour the crystal violet primary stain on it and allow to react for 1–2 minutes. Wash the slide with water to remove the excess stain.

(5) Put Gram's iodine solution on the smear and keep for about a minute.

(6) Wash the slide with 95% alcohol or acetone alcohol till the last traces of colour come out.

(7) Wash the slide with water and stain with safranin for about 30 sec.

(8) Give the final washing with water, air dry and observe under a high power microscope preferably with an oil immersion lens.

(9) The bacteria showing violet colour are gram ± ve, while the bacteria showing red or pink colour are gram –ve.

(iii) Aerobic, Facultative Anaerobic or Anaerobic :

Individual culture was inoculated in sterilised liquid Soyacasein Digest medium (Hi–Media) and by the growth pattern, differentiation was stabilished wheather they are Aerobic, Facultative anaerobic or anaerobic cultures.

(iv) Motility :

To observe motility of the organism the culture was mounted on groved slide and observed under light microscope.

(v) Fermentation :

Isolated bacteria was further checked for its fermenting nature i.e. growth with gas and acid production in Mac-Conkey's broth medium.

Procedure :

Mac-Conkey's Broth is required for this test. The composition of broth is as given below.

Bile salt or Sodium taurocholate	-	5.0 g
Peptone	-	20.0 g
Lactose	-	10.00 g
NaCl	-	5.0 g
1% Bromo cresol purple	-	1.0 ml
Distilled water	-	1.0 L
Final pH (after sterilization)	-	7.1 ± 2°C

Several dilutions of the sample for river water were prepared. Dilutions above than 1:1000 are required. 5 Fermentation tubes for each dilution, were filled with 9 ml of broth and one Durham's vial was put in each tube in inverted condition. Now all the tubes were plugged with cotton and sterilized in autoclave. All the dilutions were shaked vigorously before moving the sample to fermentation tubes. Separate sterilized pipettes were used for different dilutions as well as samples. Fermentation tubes were gently shaked. Now these tubes were incubated at 35-37°C temperature for 48 hrs. All the tubes were examined after 48 hrs for gas production.

The tube, showing gas production in Durham's tube was recorded as positive (+ ve). Even a tiny bubble should be recorded as positive. The tubes not showing gas or acid production are noted as negative (-ve).

STATISTICAL ANALYSIS OF DATA :

The data for all physico-chemical and microbiological parameters were collected and the data were analyzed for calculating mean, standard deviation of each parameter and correlation as per Goel (1999) with the help of computer on Microsoft Excel.

1. **Mean** $(\overline{x}) = \frac{\sum x}{n}$

where, $\bar{x}$ = Individual reading of parameter

$\sum x$ = Total of all readings for parameter

n = Number of samples

2. Standard Deviation (SD or SS) :

$$\text{S.D.} = \sqrt{\frac{\sum(x-\bar{x})^2}{n-1}}$$

Where, x = Individual reading of parameter

$\bar{x}$ = Mean of $\sum x$

n = Number of samples

CHAPTER-4

OBSERVATIONS

(4.a) PHYSICO-CHEMICAL

The monitoring of the river water is an essential step to mark the trend pattern of pollutants and their effect on living systems in today's developing life. The physico-chemical characteristics that were considered in the study were: -

Water temperature, conductivity, turbidity, velocity, total solids (TS), total suspended solid, total dissolved solid, pH, dissolved oxygen (DO), free CO_2, total alkalinity, biochemical oxygen demand (BOD), Chemical oxygen demand (COD), total hardness (Ca and Mg), Acidity, Chlorides, Calcium, total Nitrogen, Megnesium, Copper and Iron. Station wise details of the observations of physico-chemical parameters are mentioned in tables (Table 4a.1 to Table 4a.10) and figures (Fig.4a.1 to Fig. 4a.30)

Sampling station A:

The physico-chemical and heavy metals parameters of sampling station A are presented in Table 4a.1 to Table 4a.2 and Figures Fig. 4a.1 to Fig. 4a.6.

1. Water Temperature

The water temperature at this sampling spot was minimum 10.17°C ± 0.25°C in the December month in the year 2004-2005 and maximum of 25.50°C ± 1.26°C in March in the year 2003-2004. The average value in the year 2003-2004 was 17.55°C ± 0.44°C and 14.55°C ± 1.78°C in the year 2004-2005.

2. Conductivity

The conductivity of river water was observed to be minimum 60.18 Simens/cm ± 2.35 in the February month in the year 2003-2004 and maximum of 316.09 Simens/cm ± 0.00 in the August Month in the year 2003-2004. The average values recorded in the year 2003-2004 was 171.06 Simens/cm ± 0.03 and172.67 Simens/cm ± 0.42 in the year 2004-2005.

3. Turbidity

The turbidity of water was minimum 0.00 JTU ± 0.00 in the February month in the year 2003-2004 and maximum of 1264.00 JTU ± 13.72 in September month in the year 2004-2005. The average value of turbidity in the year 2003-2004 was 124.99 JTU ± 20.50 and 167.52 JTU ± 19.65 in the year 2004-2005.

4. Velocity

The velocity of water was minimum 0.72 m/sec ± 2.15 in the March month in the year 2004-2005 and maximum of 3.00 m/sec ± 0.00 in July month in the year 2003-2004. The average value of turbidity in the year 2003-2004 was 1.24 m/s ± 1.61 and 1.63 m/sec. ± 1.18 in the year 2004-2005.

5. Total Solids

The total solids were minimum 132.68 mg/l ± 62.50 in November month in the year 2004-2005 and maximum of 1793.68 mg/l ± 0.16 in the June month in the year 2003-2004. The average value of total solids in the year 2003-2004 was 475.94 mg/l ± 5.03 and 257.14mg/l ± 11.68 in the year 2004-2005.

6. Total Dissolved Solids

The total dissolved solids were minimum 120.07 mg/l ± 1.35 in November month in the year 2003-2004 and maximum of 895.65 mg/l ± 2.54 in the July month in the year 2004-2005. The average value of total solids in the year 2003-2004 was 205.67 mg/l ± 2.72 and 270.82 mg/l ± 3.97 in the year 2004-2005.

7. Total Suspended Solids

The total suspended solids were minimum 1.90 mg/l ± 60.86 in November month in the year 2004-2005 and maximum of 1476.08 mg/l ± 0.10 in the June month in the year 2003-2004. The average value of total solids in the year 2003-2004 was 279.37 mg/l ± 7.02 and 142.35 mg/l ± 11.41 in the year 2004-2005.

8. Potential Hydrogenii (pH)

The pH of water were observed to be minimum 6.10 ± 0.00 in the February month in the year 2003-2004 and maximum in 8.52± 1.52 in the August month in the year 2004-2005. The average value in the year 2003-2004 was 7.07 ± 0.02 and 7.66 ± 1.05 in the year 2004-2005.

9. Dissolved Oxygen (DO)

The water dissolved oxygen (DO) was minimum 7.01 mg/l ± 0.00 in the July month in the year 2004-2005 and maximum of 12.89 mg/l ± 0.03in the February month in the year 2003-2004. The average value of DO in the year 2003-2004 was observed to be 9.55 mg/l ± 0.27 and 8.90 mg/l ± 0.73 in the year 2004-2005.

10. Free Carbon dioxide (CO_2)

The free CO_2 of river water was minimum 1.15 mg/l ± 0.00 in the month of January in the year 2004-2005 and maximum of 4.50

mg/l ± 0.05 in the month of June in the year 2003-2004. The average value of free CO_2 in the year 2003-2004 was 2.85mg/l ± 0.22 and 2.50mg/l ± 0.41 in the year 2004-2005.

11. Total Alkalinity

`The total alkalinity of river water at this sampling station was minimum 25.08 mg/l ± 1.64 in the month of November in the year 2004-2005 and maximum of96.00 mg/l ± 1.65 in the month of January in the year 2004-2005. The average value in the year 2003-2005 was 82.81 mg/l ± 2.38 and 63.37 mg/l ± 1.40 in the year 2004-2005.

12. Biochemical Oxygen Demand (BOD)

The water biochemical oxygen demand (BOD) were observed to be minimum 1.65mg/l ± 0.00 in February in the year 2003-2004 and maximum 4.15 mg/l ± 1.64 in August in the year 2004-2005. The average value in the year 2003-2004 was 2.38 mg/l ± 0.26 and 3.28mg/l ± 0.56 in the year 2004-2005.

13. Chemical Oxygen Demand (COD)

The chemical oxygen demand (COD) of water was minimum 2.40 mg/l ± 0.36 in the month of June in the year 2003-2004 and maximum of 14.17 mg/l ± 0.21 in the month of January in the year 2003-2004. The average value of COD in the year 2003-2004 was 8.48 mg/l ± 0.84 and 9.06 mg/l ± 1.38 in the year 2004-2005.

14. Hardness

The total hardness of water was minimum 32.00 mg/l ± 0.16 in February month in the year 2004-2005 and maximum of 406.00 mg/l ± 0.48 in the month of August in the year 2003-2004. The average value observed in the year 2003-2004 was 93.96 mg/l ± 0.27 and 59.72 mg/l ± 2.33 in the year 2004-2005.

15. Acidity

The acidity of water was observed minimum of 36.00 mg/l ± 0.00 in October month in the year 2003-2004 and maximum of 78.00 mg/l ± 1.25 in the month of August in the year 2004-2005. The average value observed in the year 2003-2004 was 54.23 mg/l ± 0.27 and 61.01mg/l ± 1.03 in the year 2004-2005.

16. Chlorides

The Chlorides in water was minimum 6.25 mg/l ± 1.06 in the month of January in the year 2004-2005 and maximum of 24.80 mg/l ± 0.34 in the month of June in the year 2003-2004.The average value of Chlorides in the year 2003-2004 was 17.85 mg/l ± 0.20 and 13.23 mg/l ± 1.60 in the year 2004-2005.

17. Calcium

The calcium in river water at sampling station 'A' was minimum 14.25 mg/l ± 2.14 in month of May in the year 2004-2005 and maximum of 42.37 mg/l ± 0.03 in January in the year 2003-2004. The average value in the year 2003-2004 was 30.00 mg/l ± 0.40 and 23.90 mg/l ± 1.34 in the year 2004-2005.

18. Total Nitrogen

The water total nitrogen in water at this sampling station 'A' was minimum 0.14 mg/l ± 0.02 in November month in the year 2003-2004 and maximum of 2.85 mg/l ± 1.86 in the month of June in the year 2003-2004. The average value in the year 2003-2004 was 1.87 mg/l ± 0.74 and 1.94 mg/l ± 0.63 in the year 2004-2005.

19. Copper

The copper in river water at sampling station 'A' was minimum 0.02 mg/l ± 0.01 in month of November in the year 2004-2005 and maximum of 0.68 mg/l ± 0.16 in August in the year 2004-

2005. The average value in the year 2003-2004 was 0.15 mg/l ± 0.02 and 0.16mg/l ± 0.08 in the year 2004-2005.

20. Magnesium

The magnesium in river water at sampling station 'A' was minimum 13.53 mg/l ± 10.07 in month of February in the year 2003-2004 and maximum of 374.64 mg/l ± 5.30 in August in the year 2004-2005. The average value in the year 2003-2004 was 311.21 mg/l ± 2.20 and 174.30 mg/l ± 7.98 in the year 2004-2005.

21. Iron.

The concentration of iron in river water at sampling station 'A' was observed minimum of 0.15 mg/l ± 0.12 in December month in the year 2004-2005 and maximum of 0.90 mg/l ± 0.01 in the month of Jan in the year 2003-2005. The average value observed in the year 2003-2004 was 0.44 mg/l ± 0.024 and 0.31mg/l ± 0.23 in the year 2004-2005.

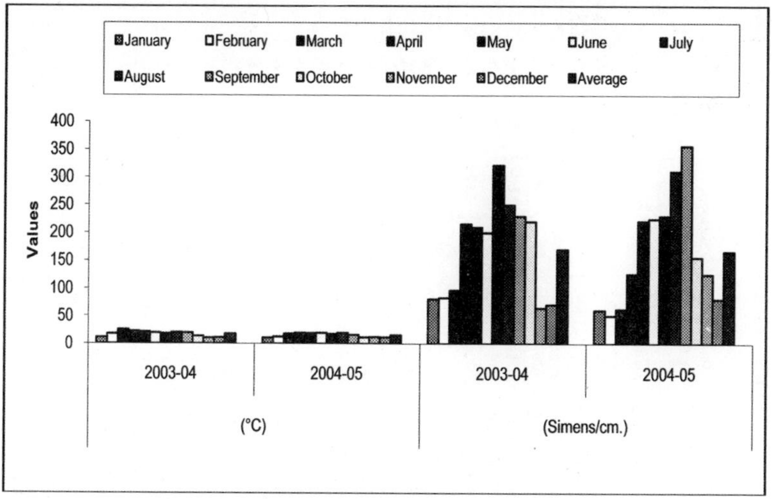

Fig. 4a.1 : Graph showing monthly fluctuation of temperature and conductivity at sampling station A

Table 4a.1: Monthly variation in Physico-chemical parameter of River Ganga During 2003-2005 at sampling station A

Months	Temperature (°C)		Conductivity (Simens/cm.)		Turbidity (J.T.U)		Velocity (m/sec.)		Total Solids (mg/l)		Total Dissolved Solids (mg/l)		Total Suspended Solids (mg/l)		PH		Dissolved Oxygen(mg/l)		Free Carbon Dioxide (mg/l)	
	2003-04	2004-05	2003-04	2004-05	2003-04	2004-05	2003-04	2004-05	2003-04	2004-05	2003-04	2004-05	2003-04	2004-05	2003-04	2004-05	2003-04	2004-05	2003-04	2004-05
January	11.02± 0.01	10.68± 5.74	92.13± 0.00	62.67± 0.86	6.85± 0.08	20.52± 0.16	1.20± 0.00	1.35± 1.24	370.00± 0.36	184.32± 0.16	189.50± 2.79	164.20± 0.21	309.64± 2.43	20.12± 0.05	6.15± 0.02	7.35± 0.15	11.50± 0.00	10.97± 0.81	1.42± 0.15	1.15± 0.00
February	16.54± 0.04	11.54± 1.16	60.18± 0.00	56.52± 1.24	0.00± 0.00	11.15± 7.16	0.74± 0.35	1.20± 0.18	367.48± 13.70	258.62± 0.78	84.20± 0.02	162.32± 0.81	283.28± 13.68	6.00± 51.19	6.10± 0.00	7.85± 2.40	12.89± 0.03	11.36± 0.80	1.81± 0.04	1.24± 0.12
March	25.50± 1.26	15.16± 2.15	80.16± 0.03	72.18± 0.26	3.26± 0.00	5.68± 2.47	0.75± 0.36	0.72± 2.15	357.17± 0.25	206.37± 1.86	135.62± 0.14	132.00± 1.80	221.55± 0.11	74.37± 0.06	7.85± 0.04	7.96± 0.20	12.35± 0.82	10.15± 1.56	4.00± 0.00	2.56± 1.18
April	21.75± 0.07	17.18± 2.15	85.20± 0.02	120.25± 0.08	42.32± 7.85	35.64± 6.32	0.85± 0.04	0.85± 1.68	389.25± 0.24	278.72± 0.36	178.52± 19.80	160.00± 1.28	190.73± 19.56	118.72± 0.89	6.18± 0.02	7.98± 0.08	9.06± 1.03	8.15± 0.16	3.30± 0.05	1.56± 0.10
May	20.25± 0.53	15.84± 0.06	162.10± 0.01	215.17± 0.24	98.62± 7.81	50.25± 8.55	2.75± 0.21	1.72± 0.00	372.71± 0.52	352.00± 3.25	193.40± 5.52	172.00± 1.27	179.31± 5.00	180.00± 1.98	7.05± 0.05	7.35± 2.54	8.12± 0.05	7.35± 0.02	3.60± 0.04	2.70± 0.00
June	20.25± 0.57	17.85± 2.14	230.12± 0.04	230.12± 0.00	346.64± 37.86	132.65± 32.52	2.90± 1.18	1.89± 1.25	1793.68± 0.16	378.62± 1.97	317.60± 0.26	214.86± 2.40	1476.08± 0.10	352.51± 0.00	7.72± 0.05	7.76± 1.20	8.58± 0.06	7.96± 1.64	4.50± 0.05	3.00± 0.00
July	20.00± 2.08	15.62± 1.18	242.48± 0.26	248.85± 0.16	435.60± 162.00	256.25± 120.00	3.00± 0.00	2.72± 1.62	1019.71± 0.39	567.37± 2.40	576.13± 0.10	895.65± 2.54	443.58± 0.29	183.46± 29.13	7.18± 0.01	7.56± 0.20	7.18± 0.12	7.01± 0.00	3.70± 0.02	2.50± 0.10
August	20.00± 0.25	18.15± 1.62	316.09± 0.00	312.12± 0.60	278.00± 3.86	115.65± 32.55	2.19± 0.02	2.60± 1.66	686.00± 7.70	677.76± 2.15	175.00± 1.80	652.08± 31.10	111.00± 5.90	115.68± 0.31	8.21± 0.02	8.52± 1.50	7.28± 0.40	7.95± 0.20	3.60± 004	3.80± 1.64
September	20.00± 0.00	15.64± 1.62	240.08± 0.02	250.64± 0.00	115.62± 7.00	1264.00± 13.72	1.95± 0.59	1.68± 0.82	193.00± 30.19	168.32± 52.00	186.71± 0.39	372.00± 2.46	22.55± 0.09	637.03± 1.76	7.69± 0.02	7.35± 2.00	8.27± 0.17	8.15± 1.16	2.60± 0.45	3.50± 0.00
October	13.56± 0.13	14.65± 1.67	235.05± 0.00	262.02± 0.00	78.00± 3.86	72.15± 5.88	1.57± 0.00	1.35± 0.15	185.00± 6.36	140.00± 11.60	135.63± 0.14	132.00± 1.85	49.37± 6.22	8.00± 9.75	7.16± 0.02	7.56± 2.14	8.59± 0.19	8.56± 0.15	2.30± 1.50	2.50± 1.64
November	10.20± 0.00	12.16± 1.60	211.07± 0.00	160.88± 0.00	56.00± 7.82	14.25± 0.80	1.99± 0.02	1.72± 1.25	179.70± 0.26	132.68± 62.50	120.07± 1.35	130.78± 1.64	5903± 1.09	1.90± 60.86	7.35± 0.01	7.87± 0.13	9.35± 0.23	8.78± 2.14	1.80± 0.12	3.50± 0.00
December	11.25± 0.35	10.17± 0.25	98.10± 0.01	80.62± 1.64	39.00± 7.86	32.04± 5.62	1.76± 0.00	1.78± 2.16	197.56± 0.18	162.35± 0.85	175.01± 0.27	152.00± 0.35	6.29± 29.80	10.35± 0.50	6.17± 0.02	6.84± 0.10	11.43± 0.08	10.35± 0.16	1.56± 0.16	1.25± 0.10
Annual Average	17.55± 0.44	14.55± 1.78	171.06± 0.03	172.67± 0.42	124.99± 20.50	167.52± 19.65	124.99± 20.50	1.63± 1.18	475.94± 5.03	257.14± 11.66	205.67± 2.72	270.82± 3.97	279.37± 7.02	142.35± 11.41	7.07± 0.02	7.66± 1.05	9.55± 0.27	8.90± 0.73	2.85± 0.22	2.50± 0.41

Sd±Standard Deviation

Table 4a.2: Monthly variation in Physico-chemical parameter of River Ganga During 2003-2005 at sampling station A

Months	Alkalinity (mg/l)		B.O.D. (mg/l)		C.O.D. (mg/l)		Hardness (mg/l)		Acidity (mg/l)		Chlorides (mg/l)		Calcium (mg/l)		Total Nitrogen (mg/l)		Copper (mg/l)		Magnesium (mg/l)		Iron (mg/l)	
	2003-04	2004-05	2003-04	2004-05	2003-04	2004-05	2003-04	2004-05	2003-04	2004-05	2003-04	2004-05	2003-04	2004-05	2003-04	2004-05	200304	2004-05	2003-04	2004-05	2003-04	2004-05
January	71.25 ±3.57	**96.00 1.65**	1.99± 0.01	3.52± 0.17	14.17± 0.21	12.00± 1.56	57.98± 1.19	48.37± 1.65	63.39± 0.32	72.00± 1.24	11.09± 0.16	6.25± 1.06	42.37± 0.03	35.08± 2.14	2.15▯ 0.15	2.10± 0.10	0.30± 0.00	0.02± 0.10	75.95± 5.64	64.66± 2.38	0.90± 0.01	0.16± 0.00
February	67.25 ±0.05	**85.64±** 1.68	1.65± 0.00	3.15± 1.68	11.53± 0.41	11.37± 1.46	42.75± 0.23	32.00± 0.16	56.00± 0.00	65.15± 1.65	19.53± 0.50	8.72± 2.48	39.97± 2.30	20.14± 1.65	2.30± 0.00	2.32± 1.46	0.04± 0.00	0.02± 0.00	13.53± 10.07	57.70± 7.25	0.19± 0.00	0.25± 0.80
March	86.25± 4.75	**72.00±** 0.24	2.25± 0.27	2.78± 0.10	8.65± 2.15	9.32± 2.03	56.50± 0.00	57.00± 1.72	46.25± 0.14	56.00± 2.54	23.15± 0.09	12.30± 1.68	35.22± 0.05	17.18± 0.30	2.72± 1.64	2.75± 1.65	0.04± 0.00	0.02± 0.17	103.54± 0.24	193.74± 6.91	0.21± 0.04	0.25± 0.80
April	88.75± 1.03	**64.36±** 2.16	2.75± 0.92	2.84± 1.65	6.82± 0.19	7.16± 1.68	55.78± 0.05	48.20± 6.12	39.38± 0.31	50.15± 1.68	23.89± 0.03	16.00± 1.77	28.75± 0.00	15.15± 0.06	2.85± 1.62	2.80± 1.15	0.05± 0.00	0.04± 0.20	131.51± 0.24	160.80± 29.48	0.24± 0.03	0.22± 0.15
May	86.25± 0.09	**85.00±** 0.14	2.35± 0.42	2.70± 0.07	16.10± 0.45	12.00± 0.50	69.00± 0.11	72.00± 8.10	65.00± 0.14	72.15± 0.11	24.75± 0.10	17.15± 1.87	24.86± 0.22	14.25± 2.14	2.60± 0.06	2.56± 1.64	0.05± 0.00	0.04± 0.00	214.76± 0.54	290.98± 29.00	0.42± 0.05	0.42± 0.12
June	75.07± 0.12	**62.15±** 0.16	2.58± 0.15	3.85± 0.80	2.40± 0.36	7.15± 1.64	59.50± 0.59	52.00± 1.00	40.62± 0.17	64.20± 0.25	24.80± 0.34	17.15± 71.62	21.64± 0.22	18.15± 1.68	2.85± 1.86	2.16± 0.00	0.57± 0.00	0.03± 0.16	184.20± 1.80	164.69± 3.31	0.56± 0.07	0.52± 0.32
July	95.17± 6.91	**25.16±** 0.04	2.98± 0.62	3.72± 0.18	5.35± 0.40	6.10± 2.14	60.00± 0.12	56.85± 2.80	47.50± 0.05	46.85± 1.68	17.62± 0.21	21.03± 0.25	20.80± 0.34	19.25± 1.62	1.37± 0.84	1.85± 0.01	0.06± 0.25	0.64± 0.00	190.72± 1.07	182.94± 5.74	0.75± 0.06	0.63± 0.32
August	74.05± 0.17	**35.15±** 1.68	2.17± 0.29	4.15± 1.64	9.05± 1.03	11.00± 0.10	406.00± 0.48	97.00± 2.15	66.25± 0.17	78.00± 1.25	15.75± 0.65	17.00± 1.54	23.85± 0.22	20.00± 1.06	1.22± 0.85	1.25± 0.10	0.10± 0.00	0.68± 0.16	1859.32± 1.27	374.64± 5.30	0.67± 0.00	0.51± 0.12
September	81.08± 0.12	**52.14±** 0.14	2.97± 0.17	3.15± 0.80	5.46± 0.34	8.10± 0.16	129.50± 0.23	75.00± 1.64	62.25± 0.12	58.16± 0.11	15.00± 0.20	15.16± 1.62	25.76± 0.15	25.12± 1.60	1.72± 0.00	1.25± 0.81	0.04± 0.00	0.05± 0.00	50.4.74± 0.39	242.69± 0.19	0.61± 0.01	0.25± 0.05
October	85.76± 4.75	**78.06±** 5.64	2.60± 0.06	3.72± 0.00	4.04± 2.00	5.17± 2.80	48.80± 0.00	54.24± 0.01	36.00± 0.00	42.00± 1.60	14.75± 0.05	12.32± 3.51	27.63± 0.06	32.15± 1.25	0.15± 0.01	1.05± 0.00	0.52± 0.02	0.05± 0.00	103.00± 0.29	107.48± 6.03	0.55± 0.00	0.21± 0.00
November	90.25± 7.39	**25.08±** 1.64	2.15± 0.13	2.15± 0.16	5.35± 2.14	7.15± 1.64	77.50± 0.17	72.00± 0.16	60.38± 1.79	52.15± 0.06	12.36± 0.00	8.56± 1.62	30.14± 1.15	38.16± 0.12	0.14± 0.02	1.02± 0.00	0.04± 0.01	0.02± 0.01	230.43± 4.77	164.65± 0.19	0.47± 0.00	0.20± 0.00
December	92.60± 0.10	**85.64±** 1.68	2.14± 0.12	3.65± 0.15	12.84± 0.44	12.15± 0.81	64.20± 0.06	52.00± 2.46	67.70± 0.03	75.35± 0.13	11.55± 0.05	7.16± 0.14	38.96± 0.08	32.15± 2.46	2.37± 1.86	2.15± 0.62	0.03± 0.00	0.35± 0.16	122.80± 0.10	96.58± 0.00	0.28± 0.02	0.15± 0.12
Annual Average	82.81± 2.38	**63.87±** 1.40	2.38± 0.26	3.28± 0.56	8.48± 0.84	9.06± 1.38	93.96± 0.27	59.72± 2.33	54.23± 0.27	61.01± 1.03	17.85± 0.20	13.23± 1.60	30.00± 0.40	23.90± 1.34	1.87± 0.74	1.94± 0.63	0.15± 0.02	0.16± 0.08	311.21± 2.20	174.30± 7.98	0.44± 0.02	0.31± 0.23

Sd±Standard Deviation

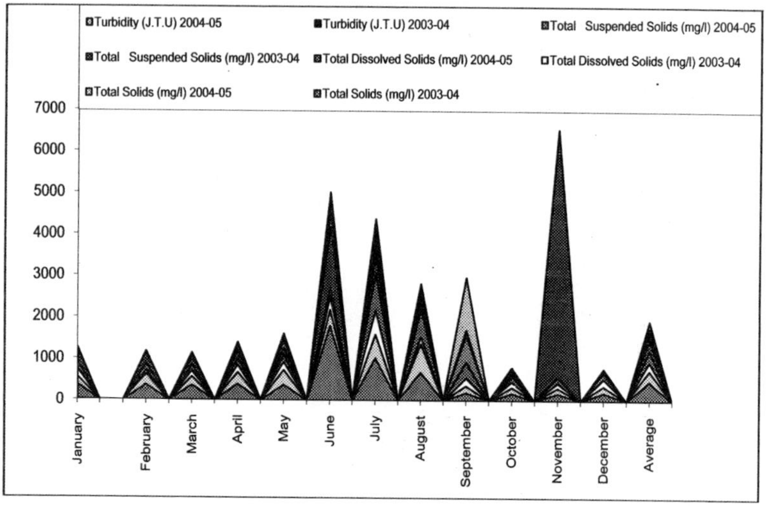

Fig. 4a.2 : Graph showing monthly fluctuation of turbidity, T.S.S., T.S. and T.D.S at Sampling station A

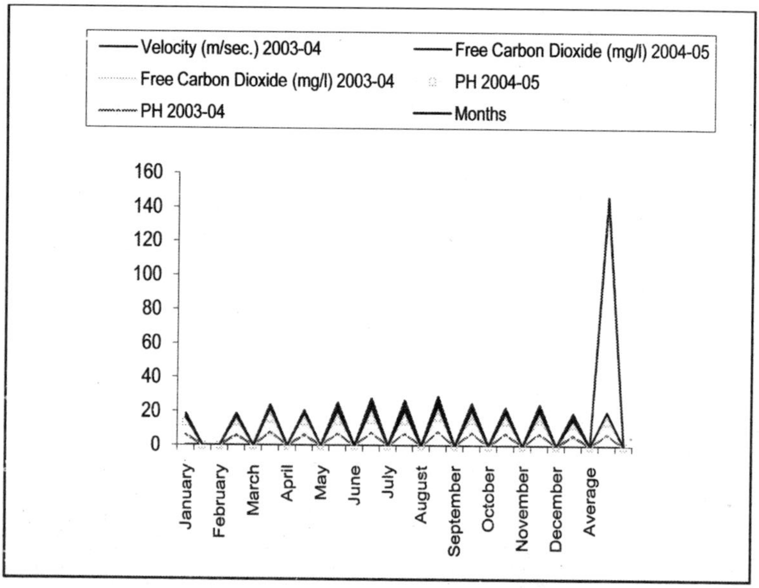

Fig. 4a.3 : Graph showing monthly fluctuation of Velocity, Free CO_2 and pH at Sampling station A

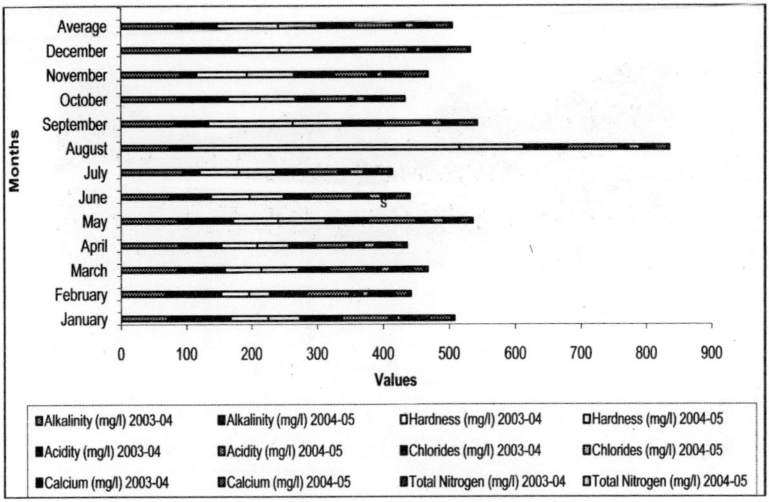

Fig. 4a.4: Graph showing monthly fluctuation of various chemical parameters at Sampling station A

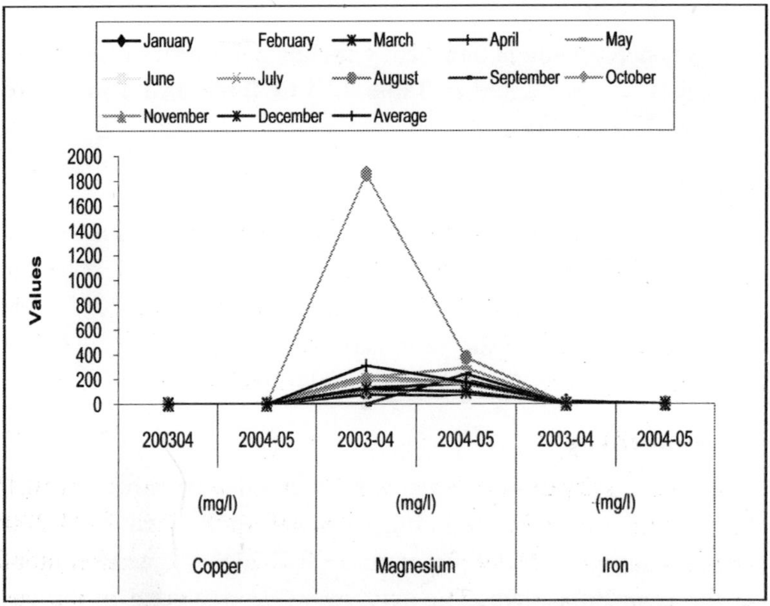

Fig. 4a.5: Graph showing monthly fluctuation of Cu, Mg and Fe at Sampling station A

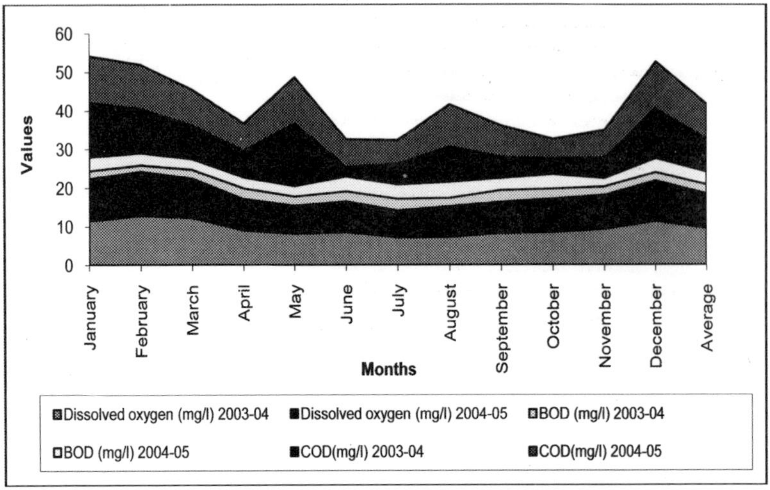

Fig. 4a.6: Graph showing monthly fluctuation of COD, BOD and DO at Sampling station A

Sampling station B: -

The physico-chemical and heavy metals parameters of sampling station B. are presented in Table 4a.3 to Table 4a.4 and Figures (Fig. 4a.7 to Fig. 4a.12).

1. Water Temperature

The water temperature at this sampling spot was minimum 10.25°C ± 0.00°C in the November month in the year 2003-2004 and maximum of 25.25°C ± 1.25°C in March in the year 2003-2004. The average value in the year 2003-2004 was 17.35°C ± 0.49°C and 15.53 °C ± 1.92°C in the year 2004-2005.

2. Conductivity

The conductivity of river water was observed to be minimum 50.18 Simens/cm ± 0.01 in the February month in the year 2004-2005 and maximum of 356.09 Simens/cm ± 0.10 in the September month in the year 2004-2005. The average values recorded in the year 2003-2004 was 170 Simens/cm ± 0.04 and 166.97 Simens/cm ± 0.50 in the year 2004-2005.

3. Turbidity

The turbidity of water was minimum 0.00 JTU ± 0.00 in the month of February and March in the year 2003-2004 and maximum of 562.64 JTU ± 178.06 in July month in the year 2003-2004. The average value of turbidity in the year 2003-2004 was 130.42 JTU ± 25.27and 67.59 JTU ± 8.6 in the year 2004-2005.

4. Velocity

The velocity of water was minimum 0.10 m/sec ± 0.00 in the March month in the year 2004-2005 and maximum of2.15 m/sec ± 0.05 in June month in the year 2003-2004. The average value of turbidity in the year 2003-2004 was 1.30 m/sec ± 0.03 and 0.90 m/sec. ± 0.07 in the year 2004-2005.

5. Total Solids

The total solids were minimum 116.22 mg/l ± 0.21 in February month in the year 2003-2004 and maximum of 1537.00 mg/l ± 0.18 in the June month in the year 2003-2004. The average value of total solids in the year 2003-2004 was 387.69 mg/l ± 0.35 and 411.19 mg/l ± 7.37 in the year 2004-2005.

6. Total Dissolved Solids

The total dissolved solids were minimum 78.21 mg/l ± 0.61 in September month in the year 2004-2005 and maximum of 324.17 mg/l ± 0.34 in the November month in the year 2003-2004. The average value of total solids in the year 2003-2004 was 161.43 mg/l ± 0.53 and 122.80 mg/l ± 12.29 in the year 2004-2005.

7. Total Suspended Solids

The total suspended solids were minimum 26.00 mg/l ± 0.2.9 in October month in the year 2004-2005 and maximum of 1142.00 mg/l ± 17.55 in the July month in the year 2004-2005. The average value of total solids in the year 2003-2004 was 187.56 mg/l ± 5.47 and 288.39 mg/l ± 9.72 in the year 2004-2005.

8. Potential Hydrogenii (pH)

The pH of water were observed to be minimum 6.13 ± 0.01 in the December month in the year 2003-2004 and maximum in 8.85 ± 0.67 in the August month in the year 2004-2005. The average value in the year 2003-2004 was 7.11 ± 0.03 and 7.60 ± 0.13 in the year 2004-2005.

9. Dissolved Oxygen (DO)

The water dissolved oxygen (DO) was minimum 8.12 mg/l ± 0.10 in the August month in the year 2004-2005 and maximum of 13.15 mg/l ± 0.03 in the February month in the year 2003-2004. The average value of DO in the year 2003-2004 was observed to be 9.73 mg/l ± 0.13 and 9.28 mg/l ± 0.14 in the year 2004-2005.

10. Free Carbon Dioxide (CO_2)

The free CO_2 of river water was minimum 1.15 mg/l ± 0.01 in the month of September in the year 2003-2004 and maximum of 4.50 mg/l ± 0.05 in the month of May in the year 2003-2004. The average value of free CO_2 in the year 2003-2004 was 2.37 mg/l ± 0.09 and 2.58 mg/l ± 0.11 in the year 2004-2005.

11. Total Alkalinity

The total alkalinity of river water at this sampling station was minimum 44.07 mg/l ± 3.55 in the month of February in the year 2003-2004 and maximum of 87.01 mg/l ± 1.47 in the month of November in the year 2003-2004. The average value in the year 2003-2005 was 70.92 mg/l ± 0.91 and 75.39 mg/l ± 7.97 in the year 2004-2005.

12. Biochemical Oxygen Demand (BOD)

The water biochemical oxygen demand (BOD) were observed to be minimum 1.75 mg/l ± 0.06 in February in the year 2003-2004 and maximum 5.64 mg/l ± 0.21 in July in the year 2004-2005.

The average value in the year 2003-2004 was 2.63 mg/l ± 0.20 and 3.15 mg/l ± 0.14 in the year 2004-2005.

13. Chemical Oxygen Demand (COD)

The chemical oxygen demand (COD) of water was minimum 3.16 mg/l ± 0.24 in the month of August in the year 2004-2005 and maximum of 44.00 mg/l ± 0.26 in the month of June in the year 2003-2004. The average value of COD in the year 2003-2004 was 11.86 mg/l ± 0.77 and 7.26 mg/l ± 1.19 in the year 2004-2005.

14. Hardness

The total hardness of water was minimum 27.50 mg/l ± 0.00 in March month in the year 2003-2004 and maximum of 131.00 mg/l ± 0.23 in the month of September in the year 2003-2004. The average value observed in the year 2003-2004 was 65.05 mg/l ± 0.15 and 68.85 mg/l ± 4.88 in the year 2004-2005.

15. Acidity

The acidity of water was observed minimum of 18.50 mg/l ± 0.10 in July month in the year 2003-2004 and maximum of 66.24 mg/l ± 0.06 in the month of Mach in the year 2003-2004.The average value observed in the year 2003-2004 was 48.13 mg/l ± 0.07 and 51.29 mg/l ± 2.19 in the year 2004-2005.

16. Chlorides

The Chlorides in water was minimum 8.61 mg/l ± 7.10 in the month of January in the year 2004-2005 and maximum of 28.35 mg/l ± 15.00 in the month of July in the year 2004-2005.The average value of Chlorides in the year 2003-2004 was 18.49 mg/l ± 0.15 and 15.09 mg/l ± 4.70 in the year 2004-2005.

17. Calcium

The calcium in river water at sampling station 'B ' was minimum 16.78 mg/l ± 0.50 in month of July in the year 2003-2004 and maximum of 47.00 mg/l ± 0.07 in January in the year 2003-2004. The average value in the year 2003-2004 was 29.10 mg/l ± 0.21 and 26.66 mg/l ± 2.04 in the year 2004-2005.

18. Total Nitrogen

The water total nitrogen in water at this sampling station 'B' was minimum 0.20 mg/l ± 0.00 in November month in the year 2004-2005 and maximum of 2.87 mg/l ± 0.30 in the month of April in the year 2004-2005. The average value in the year 2003-2004 was 1.79 mg/l ± 0.65 and 1.89 mg/l ± 0.39 in the year 2004-2005.

19. Copper

The copper in river water at sampling station 'B' was minimum 0.02 mg/l ± 0.00 in month of November in the year 2004-2005 and maximum of 0.37 mg/l ± 0.04 in December in the year 2003-2004. The average value in the year 2003-2004 was 0.10 mg/l ± 0.03 and 0.03 mg/l ± 0.03 in the year 2004-2005.

20. Magnesium

The Magnesium in river water at sampling station 'B' was minimum 27.83 mg/l ± 0.15 in month of March in the year 2003-2004 and maximum of 543.03 mg/l ± 3.16 in September in the year 2003-2004.The average value in the year 2003-2004 was 174.90 mg/l ± 0.87 and 205.30 mg/l ± 18.96 in the year 2004-2005.

21. Iron

The concentration of iron in river water at sampling station 'B' was observed minimum of 0.20 mg/l ± 0.01 in January month in the year 2003-2004 and maximum of 0.93 mg/l ± 0.03 in the month of July in the year 2003-2004. The average value observed in the year 2003-2004 was 0.49 mg/l ± 0.05 and 0.43 mg/l ± 0.21 in the year 2004-2005.

Table 4a.3: Monthly variation in Physico-chemical parameter of River Ganga During 2003-2005 at sampling station B

Months	Temperature (°C)		Conductivity (Simens/cm.)		Turbidity (J.T.U)		Velocity (m/sec.)		Tatal Solids (mg/l)		Total Dissolved Solids (mg/l)		Total Suspended Solids (mg/l)		PH		Dissolved Oxygen (mg/l)		Free Carbon Dioxide (mg/l)	
	2003-04	2004-05	2003-04	2004-05	2003-04	2004-05	2003-04	2004-05	2003-04	2004-05	2003-04	2004-05	2003-04	2004-05	2003-04	2004-05	2003-04	2004-05	2003-04	2004-05
January	11.20± 0.00	10.57± 0.32	80.12± 0.03	60.35± 0.02	12.00± 2.06	16.72± 1.25	1.35± 0.01	0.26± 0.12	138.00± 0.34	335.06± 0.25	87.51± 0.04	98.71± 3.54	58.78± 0.02	236.35± 3.29	7.17± 0.00	7.30± 5.00	11.21± 0.00	11.26± 0.05	1.35± 0.03	1.32± 0.01
February	17.51± 0.25	12.52± 0.35	82.23± 0.10	50.18± 0.01	0.00± 0.00	9.10± 0.00	0.21± 0.02	0.17± 0.02	116.22± 0.21	368.00± 25.50	98.90± 0.00	86.25± 0.03	49.77± 0.06	281.75± 25.47	7.21± 0.02	7.85± 0.05	13.15± 0.03	10.32± 0.05	1.20± 0.00	1.60± 0.03
March	25.25± 1.25	17.85± 0.05	96.14± 0.02	62.19± 0.12	0.00± 0.00	5.26± 0.05	0.39± 0.00	0.10± 0.00	135.00± 0.04	372.56± 2.35	85.23± 0.10	86.27± 0.21	180.10± 0.06	284.29± 2.14	6.40± 0.11	6.25± 0.02	12.42± 0.02	10.10± 0.20	1.37± 0.06	1.54± 0.10
April	22.25± 1.05	19.35± 0.20	216.02± 0.00	125.25± 0.10	42.65± 7.86	15.68± 2.15	0.45± 0.05	0.56± 0.05	235.00± 1.46	368.00± 0.25	95.01± 1.52	136.00± 25.01	139.99± 25.12	232.00± 24.76	6.28± 0.02	7.81± 0.01	9.10± 0.03	9.60± 0.02	3.15± 0.03	3.78± 0.60
May	20.75± 0.50	19.00± 0.38	210.10± 0.00	221.10± 0.00	97.98± 6.81	68.52± 6.34	2.07± 0.02	1.06± 0.00	273.02± 0.58	352.00± 11.50	104.27± 0.24	104.00± 10.30	168.75± 19.41	248.00± 1.20	7.11± 0.00	7.56± 0.02	8.18± 0.17	8.54± 0.05	4.50± 0.05	2.70± 0.30
June	19.25± 0.37	19.35± 0.05	200.12± 0.01	225.15± 0.03	346.51± 62.00	98.00± 14.64	2.15± 0.05	1.35± 0.20	1537.00± 0.18	1036.00± 3.52	128.00± 1.16	95.00± 15.16	140.90± 13.88	783.47± 58.56	7.60± 0.04	7.82± 0.06	8.20± 0.24	6.20± 0.20	3.85± 0.27	2.30± 0.02
July	17.75± 0.23	17.23± 3.90	321.48± 0.15	230.52± 0.35	562.64± 178.06	36.15± 18.27	1.77± 0.03	1.20± 0.00	1160.00± 0.24	1237.00± 32.51	229.04± 0.86	252.53± 62.08	930.96± 0.62	1142.00± 17.55	7.13± 0.01	7.35± 0.12	8.68± 0.07	6.86± 0.16	3.50± 0.03	4.20± 0.02
August	19.75± 0.50	19.25± 2.12	250.09± 0.03	310.51± 5.20	245.00± 3.00	242.00± 30.25	1.78± 0.03	1.76± 0.25	574.01± 0.06	786.25± 0.32	175.00± 0.18	135.00± 5.64	147.50± 0.05	51.25± 5.32	8.20± 0.08	8.85± 0.67	8.97± 0.09	8.12± 0.10	1.35± 0.49	3.60± 0.01
September	19.50± 0.50	15.68± 15.00	230.10± 0.01	356.09± 0.10	172.62± 27.00	157.00± 18.35	1.65± 0.03	1.78± 0.03	235.00± 0.06	142.00± 0.36	221.00± 1.80	78.21± 0.61	172.79± 0.15	63.79± 0.25	7.55± 0.01	7.80± 0.05	8.99± 0.26	8.56± 0.35	1.15± 0.01	3.10± 0.00
October	13.52± 1.06	11.73± 0.25	220.07± 0.03	156.08± 0.00	30.00± 2.80	78.00± 2.65	1.15± 0.01	1.85± 0.02	196.00± 0.20	168.00± 1.35	225.03± 0.00	142.00± 4.26	128.17± 0.64	26.00± 2.91	7.17± 0.00	7.57± 0.12	8.17± 0.53	10.35± 0.20	2.54± 0.02	2.50± 0.02
November	10.25± 0.00	12.21± 0.16	64.18± 0.03	125.03± 0.00	15.67± 7.35	32.00± 5.00	1.10± 0.00	0.35± 0.02	175.00± 0.60	187.00± 10.30	324.17± 0.34	97.26± 18.05	83.00± 0.31	89.74± 7.75	7.35± 0.00	8.16± 0.02	8.37± 0.07	9.25± 0.03	1.39± 0.09	2.70± 0.16
December	11.25± 0.12	11.65± 0.26	70.15± 0.05	80.16± 0.03	40.00± 6.24	52.60± 4.32	1.55± 0.05	0.32± 0.14	173.00± 0.21	182.35± 0.18	164.00± 0.06	160.35± 2.56	50.00± 6.62	22.00± 2.38	6.13± 0.01	6.85± 0.12	11.32± 0.03	12.21± 0.32	1.39± 0.09	1.65± 0.02
Annual Average	17.35± 0.49	15.53± 1.92	170.07± 0.04	166.97± 0.50	130.42± 25.27	67.59± 8.61	1.30± 0.03	0.90± 0.07	387.69± 0.35	411.19± 7.37	161.43± 0.53	122.80± 12.29	187.56± 5.47	288.39± 9.72	7.11± 0.03	7.60± 0.52	9.73± 0.13	9.28± 0.14	2.37± 0.09	2.58± 0.11

Sd±Standard Deviation

Table 4a.4: Monthly variation in Physico-chemical parameter of River Ganga During 2003-2005 at sampling station B

Months	Alkalinity (mg/l)		B.O.D. (mg/l)		C.O.D. (mg/l)		Hardness (mg/l)		Acidity (mg/l)		Chlorides (mg/l)		Calcium (mg/l)		Total Nitrogen (mg/l)		Copper (mg/l)		Magnesium (mg/l)		Iron (mg/l)	
	2003-04	2004-05	2003-04	2004-05	2003-04	2004-05	2003-04	2004-05	2003-04	2004-05	2003-04	2004-05	2003-04	2004-05	2003-04	2004-05	2003-04	2004-05	2003-04	2004-05	2003-04	2004-05
January	72.17± 2.25	86.52± 5.64	2.19± 0.01	2.58± 0.02	14.15± 0.20	12.36± 5.24	69.34± 0.02	56.52± 10.35	55.15± 0.05	52.51± 0.24	11.78± 0.03	8.61± 7.10	47.00± 0.07	35.20± 0.21	2.06± 0.10	2.15± 0.01	0.03± 0.00	0.04± 0.05	108.69± 0.24	103.73♀ 48.85	0.20± 0.01	0.35± 0.03
February	44.07± 3.55	78.06± 0.12	1.75± 0.06	2.32± 0.06	11.71± 0.42	10.32± 0.20	51.50± 0.09	42.16± 0.25	46.71± 0.03	50.00± 8.00	19.53± 0.00	19.68± 8.07	38.92± 0.09	30.08± 0.10	2.32± 1.87	2.35± 1.68	0.36± 0.00	0.05± 0.05	39.75± 0.39	58.77♀ 0.73	0.32± 0.12	0.26± 0.03
March	85.00± 0.06	85.64± 0.68	2.37± 0.17	2.15± 0.01	8.76± 3.14	9.62± 2.50	27.50± 0.00	68.17± 8.20	66.24± 0.06	47.00± 0.30	20.00± 0.00	10.30± 2.00	33.22± 0.03	26.52± 2.50	2.52± 1.06	2.56± 1.36	0.03± 0.01	0.03± 0.06	27.83± 0.15	202.64♀ 27.73	0.40± 0.04	0.25± 0.00
April	81.25± 0.09	96.35± 2.16	2.56± 0.02	2.12± 0.00	6.71± 0.17	7.85± 0.16	57.00± 0.05	72.88± 8.61	46.87± 0.02	45.26± 3.02	21.05± 0.06	15.10± 6.10	30.02± 0.00	18.35± 8.01	2.72± 0.86	2.87± 0.30	0.04± 0.00	0.03± 0.00	131.27± 0.24	265.31♀ 2.92	0.42± 0.02	0.35± 0.16
May	75.00± 0.00	85.00± 10.25	2.77± 0.08	3.62± 0.07	15.80± 0.64	9.26± 0.25	30.75± 0.01	65.25± 3.56	65.00± 0.17	65.36± 0.31	22.31± 0.05	21.00± 0.15	27.07± 0.21	18.00± 0.02	2.56± 0.85	2.56± 0.02	0.05± 0.00	0.04± 0.00	118.86± 0.58	229.89♀ 17.22	0.47± 0.01	0.43± 0.07
June	68.17± 0.12	78.26± 3.25	2.93± 0.48	4.26± 0.52	44.00± 0.26	3.68± 0.23	67.50± 0.40	52.51± 0.26	37.50± 0.07	42.52± 0.24	23.94± 0.30	18.62± 6.25	25.62± 0.52	17.35± 6.10	2.76± 1.06	2.35± 1.25	0.06± 0.00	0.03± 0.00	203.76± 0.58	171.07± 28.41	0.57± 0.17	0.52± 0.26
July	55.52± 2.23	68.35± 0.26	3.71± 0.45	5.64± 0.21	5.78± 0.37	5.26± 0.25	53.57± 0.09	72.35± 3.08	18.50± 0.10	28.36± 0.25	23.19± 0.21	28.35± 15.00	16.78± 0.50	17.26± 0.24	1.25± 0.24	1.20± 0.00	0.08± 0.25	0.05± 0.05	179.00♀ 1.99	268.04± 13.82	0.93± 0.03	0.65± 0.26
August	73.75± 0.01	55.21± 0.35	2.75± 0.23	4.21± 0.32	9.10± 0.40	3.16± 0.24	100.00± 0.71	96.52± 8.26	46.75± 0.08	65.00± 6.00	20.51± 0.75	15.30± 2.50	18.05± 0.12	25.62± 3.16	1.15± 0.00	1.16± 0.01	0.05± 0.00	0.05± 0.04	398.72♀ 2.87	344.96± 24.81	0.73± 0.03	0.52± 0.30
September	56.79± 0.85	48.26± 0.51	2.82± 0.37	3.50± 0.20	5.30± 0.12	4.26± 0.20	131.00± 0.23	104.00± 12.56	46.75± 0.17	52.50± 0.30	17.50± 0.20	12.10± 6.08	19.39± 0.88	26.35± 0.25	1.62± 0.87	1.65± 0.04	0.04± 0.00	0.03± 0.03	543.03♀ 3.16	377.80± 59.89	0.67± 0.10	0.56± 0.62
October	75.24± 0.27	85.68± 30.15	2.71± 0.31	2.58± 0.02	5.14± 2.01	5.35± 2.16	44.30± 0.02	56.80± 0.21	34.50± 0.00	48.25± 7.06	16.23± 0.03	11.35± 0.24	27.63± 0.06	35.00± 0.16	0.17± 0.00	1.25± 0.02	0.04± 0.02	0.02± 0.08	81.11♀ 0.19	106.07± 0.24	0.58± 0.01	0.47± 0.04
November	87.01± 1.47	65.36± 7.26	2.69± 0.19	2.50± 0.02	3.30± 1.10	4.65± 2.60	77.00± 0.07	76.26± 3.01	55.57± 0.05	56.25± 0.26	13.62± 0.02	10.32± 2.60	30.34± 0.04	37.51± 3.52	0.12± 0.02	0.20± 0.00	0.02± 0.00	0.02± 0.00	227.02♀ 0.15	188.53± 2.48	0.45± 0.01	0.33± 0.12
December	77.10± 0.00	72.00± 35.01	2.35± 0.06	2.37± 0.20	12.57± 0.35	11.35± 0.21	71.10± 0.10	62.80± 0.35	58.00± 0.02	62.51+ 0.26	12.26± 0.11	10.36± 0.25	35.15± 0.05	32.62± 0.26	2.25± 0.86	2.32± 0.02	0.37± 0.04	0.03± 0.02	174.91♀ 0.24	146.84± 0.44	0.21± 0.00	0.52± 0.35
Annual Average	70.92± 0.91	75.39± 7.97	2.63± 0.20	3.15± 0.14	11.86± 0.77	7.26± 1.19	65.05± 0.15	68.85± 4.88	48.13± 0.07	51.29± 2.19	18.49± 0.15	15.09± 4.70	29.10± 0.21	26.66± 2.04	1.79± 0.65	1.89± 0.39	0.10± 0.03	0.03± 0.03	174.90♀ 0.87	205.30± 18.96	0.49± 0.05	0.43± 0.21

Sd±Standard Deviation

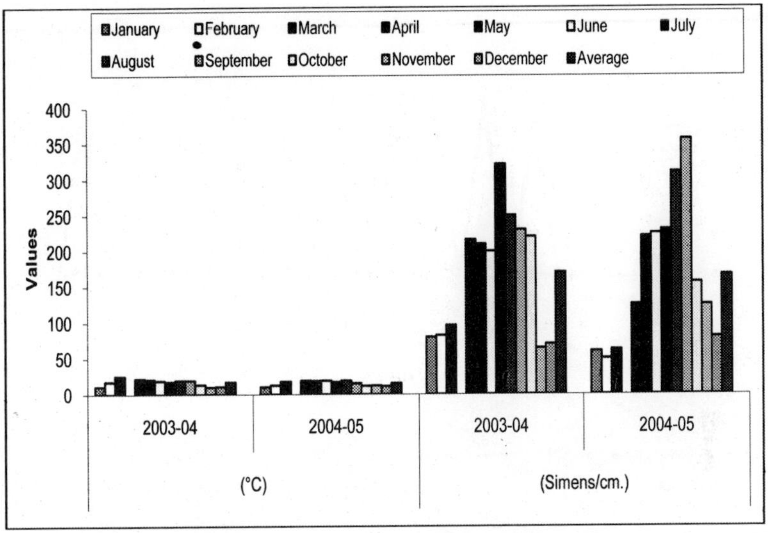

Fig. 4a.7 : Graph showing monthly fluctuation of temperature and conductivity at sampling station B

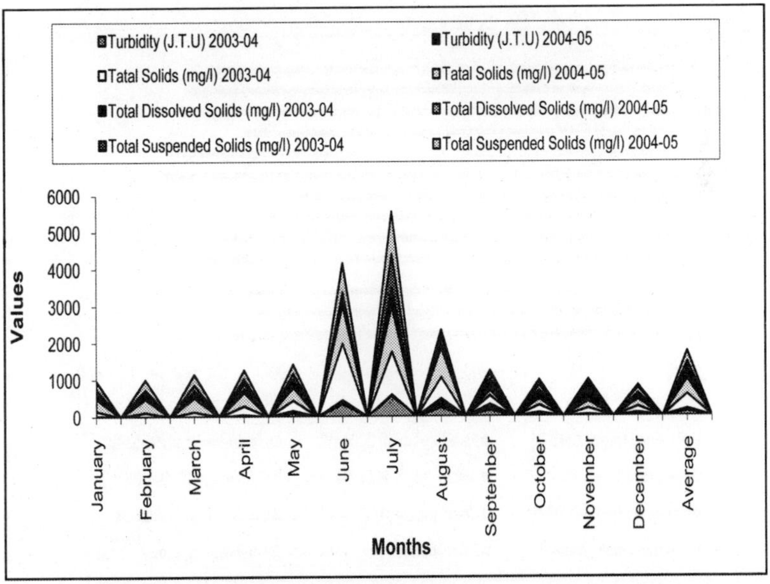

Fig. 4a.8 : Graph showing monthly fluctuation of turbidity, T.S.S., T.S. and T.D.S at Sampling station B

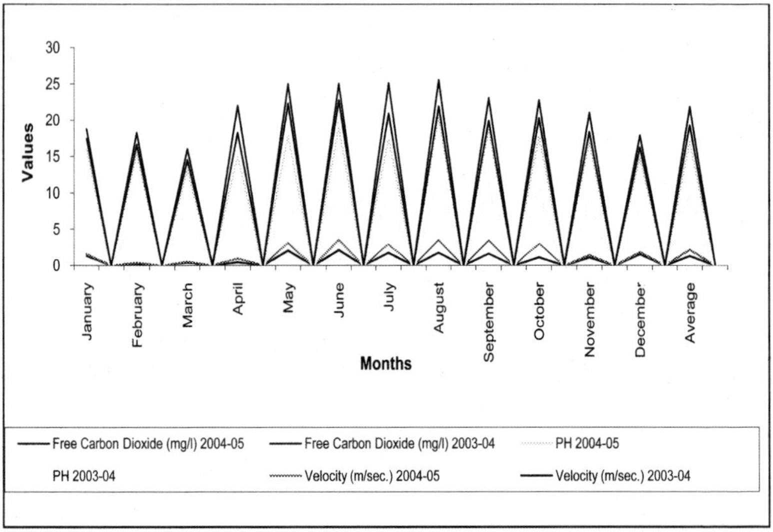

Fig. 4a.9 : Graph showing monthly fluctuation of Velocity, Free CO_2 and pH at Sampling station B

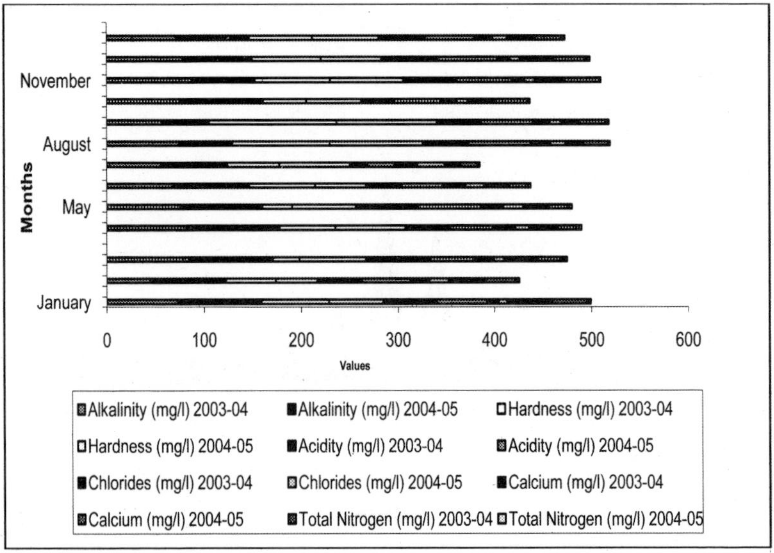

Fig. 4a.10: Graph showing monthly fluctuation of various chemical parameters at Sampling station B

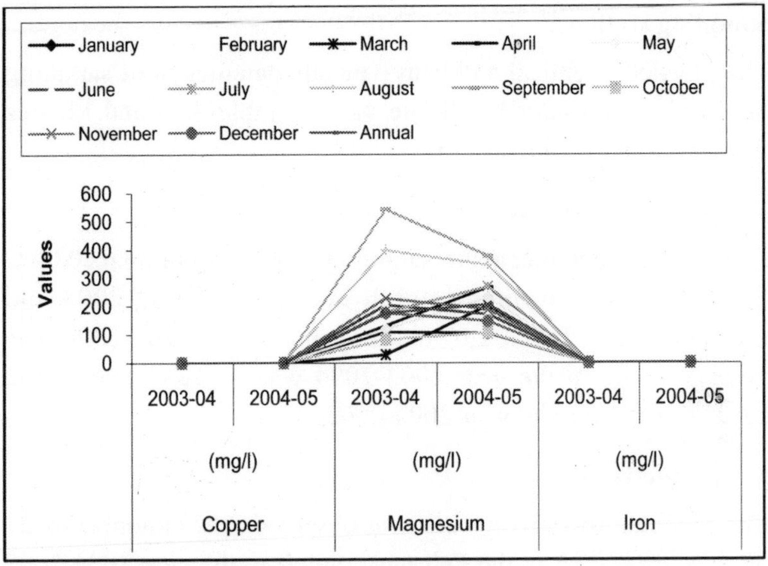

Fig. 4a.11: Graph showing monthly fluctuation of Cu, Mg and Fe at Sampling station B

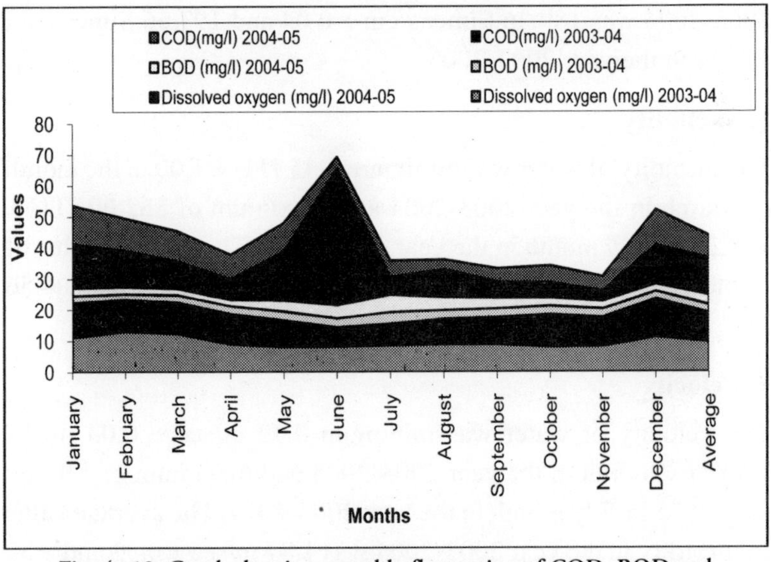

Fig. 4a.12: Graph showing monthly fluctuation of COD, BOD and DO at Sampling station B

Sampling station C

The physico-chemical and heavy metals parameters of sampling sation C are presented in Table 4a.5 to Table 4a.6 and Figures (Fig.4a.13 to Fig. 4a.18).

1. Water Temperature

The water temperature at this sampling spot was minimum 10.00°C ± 0.00 in the November month in the year 2003-2004 and maximum of 25.75°C ± 1.07 in March in the year 2003-2004. The average value in the year 2003-2004 was 17.38°C ± 0.48 and 15.67°C ± 0.40 in the year 2004-2005.

2. Conductivity

The conductivity of river water was observed to be minimum 60.21 Simens/cm ± 0.16 in the February month in the year 2004-2005 and maximum of 326.03 Simens/cm ± 0.03 in the Augest month in the year 2004-2005. The average values recorded in the year 2003-2004 was 180.36 Simens/cm ± 0.04 and 197.66 Simens/cm ± 0.06 in the year 2004-2005.

3. Turbidity

The turbidity of water was minimum 2.45 JTU ± 1.00 in the month of March in the year 2003-2004 and maximum of 362.00 JTU ± 158.25 in July month in the year 2004-2005. The average value of turbidity in the year 2003-2004 was 90.44 JTU ± 19.01 and 104.56 JTU ± 21.58 in the year 2004-2005.

4. Velocity

The velocity of water was minimum 0.32 m/sec ± 0.03 in the February month in the year 2004-2005 and maximum of 3.90 m/sec ± 0.75 in July month in the year 2003-2004. The average value of turbidity in the year 2003-2004 was 1.87 m/sec ± 0.52 and 1.31 m/sec. ± 0.03 in the year 2004-2005.

5. Total Solids

The total solids were minimum136.00 mg/l ± 0.12 in September month in the year 2004-2005 and maximum of 1321.00 mg/l ± 0.16 in the June month in the year 2004-2005. The average value of total solids in the year 2003-2004 was 438.89 mg/l ± 11.10 and 372.52 mg/l ± 0.50 in the year 2004-2005.

6. Total Dissolved Solids

The total dissolved solids were minimum 62.15 mg/l ± 0.20 in March month in the year 2004-2005 and maximum of 456.00 mg/l ± 21.10 in the July month in the year 2003-2004. The average value of total solids in the year 2003-2004 was 203.32 mg/l ± 2.43 and 156.74 mg/l ± 0.66 in the year 2004-2005.

7. Total Suspended Solids

The total suspended solids were minimum 3.05 mg/l ± 1.90 in December month in the year 2004-2005 and maximum of 1194.85 mg/l ± 1.09 in the July month in the year 2004-2005. The average value of total solids in the year 2003-2004 was 235.57 mg/l ± 12.87 and 265.65 mg/l ± 0.55 in the year 2004-2005.

8. Potential Hydrogenii (pH)

The pH of water were observed to be minimum 6.11 ± 0.02 in the December month in the year 2003-2004 and maximum in 8.75 ± 0.54 in the August month in the year 2003-2004. The average value in the year 2003-2004 was 7.66 ± 0.07 and 7.46 ± 0.17 in the year 2004-2005.

9. Dissolved Oxygen (DO)

The water dissolved oxygen (DO) was minimum 7.58 mg/l ± 0.23 in the July month in the year 2003-2004 and maximum of 12.35 mg/l ± 0.02 in the February month in the year 2004-2005. The average value of DO in the year 2003-2004 was observed to be 9.61 mg/l ± 0.24 and 9.57 mg/l ± 0.08 in the year 2004-2005.

10. Free Carbon Dioxide (CO_2)

The free CO_2 of river water was minimum 1.13 mg/l ± 0.00 in the month of February in the year 2003-2004 and maximum of 4.80 mg/l ± 0.00 in the month of June in the year 2003-2004. The average value of free CO_2 in the year 2003-2004 was 2.89 mg/l ± 0.11 and 2.50 mg/l ± 0.05 in the year 2004-2005.

11. Alkalinity

The alkalinity of river water at this sampling station was minimum 52.00 mg/l ± 0.40 in the month of February in the year 2003-2004 and maximum of 104.00 mg/l ± 8.97 in the month of January in the year 2003-2004. The average value in the year 2003-2005 was 81.35 mg/l ± 2.17 and 82.60 mg/l ± 12.99 in the year 2004-2005.

12. Biochemical Oxygen Demand (BOD)

The water biochemical oxygen demand (BOD) were observed to be minimum 1.17 mg/l ± 0.00 in February in the year 2003-2004 and maximum 3.71 mg/l ± 0.12 in September in the year 2003-2004. The average value in the year 2003-2004 was 2.59 mg/l ± 0.20 and 2.46 mg/l ± 0.19 in the year 2004-2005.

13. Chemical Oxygen Demand (COD)

The chemical oxygen demand (COD) of water was minimum 3.50mg/l ± 0.80 in the month of June in the year 2004-2005 and maximum of 15.58 mg/l ± 0.51 in the month of January in the year 2004-2005. The average value of COD in the year 2003-2004 was 8.64 mg/l ± 1.15 and 8.44 mg/l ± 0.85 in the year 2004-2005.

14. Hardness

The total hardness of water was minimum 40.25 mg/l ± 0.05 in February month in the year 2003-2004 and maximum of 326.72 mg/l ± 0.16 in the month of August in the year 2004-2005. The average value observed in the year 2003-2004 was 66.42 mg/l ± 0.36 and 91.08 mg/l ± 0.70 in the year 2004-2005.

15. Acidity

The acidity of water was observed minimum of 32.25 mg/l ± 0.26 in April month in the year 2004-2005 and maximum of 89.55 mg/l ± 2.52 in the month of January in the year 2003-2004.The average value observed in the year 2003-2004 was 56.88 mg/l ± 0.58 and 51.44 mg/l ± 0.63 in the year 2004-2005.

16. Chlorides

The Chlorides in water was minimum 5.71 mg/l ± 6.13 in the month of January in the year 2003-2004 and maximum of 28.00 mg/l ± 0.00 in the month of July in the year 2004-2005.The average value of Chlorides in the year 2003-2004 was 12.76 mg/l ± 1.57 and 18.41 mg/l ± 0.18 in the year 2004-2005.

17. Calcium

The calcium in riverwater at sampling station 'C ' was minimum 18.05 mg/l ± 0.53 in month of July in the year 2003-2004 and maximum of 42.00 mg/l ± 0.05 in March in the year 2004-2005. The average value in the year 2003-2004 was 27.47 mg/l ± 1.20 and 29.20 mg/l ± 0.51 in the year 2004-2005.

18. Total Nitrogen

The water total nitrogen in water at this sampling station 'C' was minimum 0.12mg/l ± 0.01 in November month in the year 2004-2005 and maximum of 2.82 mg/l ± 1.62 in the month of June in the year 2004-2005. The average value in the year 2003-2004 was 1.73 mg/l ± 0.36 and 1.79 mg/l ± 0.59 in the year 2004-2005.

19. Copper

The copper in riverwater at sampling station 'C' was minimum 0.01 mg/l ± 0.01 in month of December in the year 2004-2005 and maximum of 0.62 mg/l ± 0.00 in September in the year 2003-2004. The average value in the year 2003-2004 was 0.13 mg/l ± 0.03 and 0.09 mg/l ± 0.04 in the year 2004-2005.

20. Magnesium

The Magnesium in riverwater at sampling station 'C' was minimum 27.98 mg/l ± 0.24 in month of February in the year 2003-2004 and maximum of 1467.26 mg/l ± 0.19 in August in the year 2004-2005. The average value in the year 2003-2004 was 189.55 mg/l ± 5.93 and 301.09mg/l ± 5.04 in the year 2004-2005.

21. Iron

The concentration of iron in riverwater at sampling station 'C' was observed minimum of 0.16 mg/l ± 0.00 in February month in the year 2004-2005 and maximum of 0.73 mg/l ± 0.03 in the month of July in the year 2003-2004. The average value observed in the year 2003-2004 was 0.43 mg/l ± 0.03 and 0.42 mg/l ± 0.02 in the year 2004-2005.

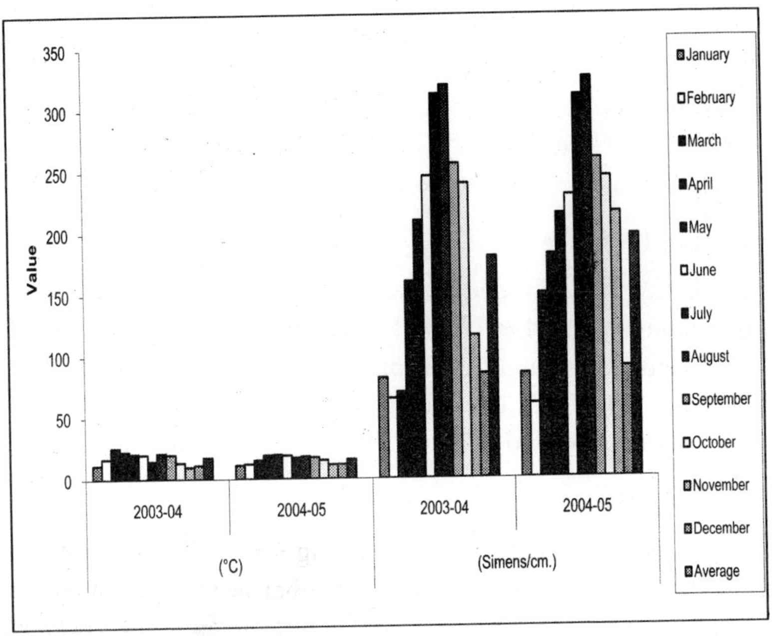

Fig. 4a.13 : Graph showing monthly fluctuation of temperature and conductivity at sampling station C

Table 4a.5: Monthly variation in Physico-chemical parameter of River Ganga During 2003-2005 at sampling station C

Months	Temperature (°C)		Conductivity (Simens/cm.)		Turbidity (J.T.U)		Velocity (m/sec.)		Tatal Solids (mg/l)		Total Dissolved Solids (mg/l)		Total Suspended Solids (mg/l)		PH		Dissolved Oxygen (mg/l)		Free Carbon Dioxide (mg/l)	
	2003-04	2004-05	2003-04	2004-05	2003-04	2004-05	2003-04	2004-05	2003-04	2004-05	2003-04	2004-05	2003-04	2004-05	2003-04	2004-05	2003-04	2004-05	2003-04	2004-05
January	11.50± 0.01	11.20 ±0.0	82.13 ±0.01	85.10± 0.00	12.37 ±0.86	20.15 ±2.10	1.56 ±1.15	1.50 ±0.01	279.36 ±0.18	232.00 ±0.06	178.50 ±0.43	96.25 ±0.12	100.76± 0.25	135.75 ±0.06	7.19 ±0.01	7.25 ±0.12	11.75 ±0.05	10.01 ±0.00	1.35 ±0.01	1.15 ±0.01
February	16.75± 0.06	12.15 ±0.15	65.15 ±0.05	60.21± 0.16	8.00 ±0.00	8.10 ±0.02	0.87 ±0.31	0.32 ±0.03	374.60 ±0.35	248.05 ±0.09	88.00 ±0.03	85.00 ±0.08	286.80± 0.32	163.05 ±0.01	7.27 ±0.06	7.12 ±0.16	12.07± 0.00	12.35 ±0.02	1.13 ±0.00	1.16 ±0.00
March	25.75± 1.07	15.25 ±0.56	70.21 ±0.02	150.15± 0.05	2.45 ±1.00	5.06 ±0.05	1.23 ±0.30	0.35 ±0.05	335.00 ±0.19	156.00 ±0.03	162.71 ±0.10	62.15 ±0.20	172.29± 0.09	93.85 ±0.18	7.74 ±0.03	6.58 ±0.12	11.01± 0.57	11.52 ±0.03	4.30 ±0.16	1.26 ±0.02
April	22.50± 1.05	19.25 ±1.06	160.17 ±0.00	182.16± 0.02	30.62 ±7.81	25.15 ±6.12	0.93 ±0.19	0.50 ±0.00	351.04 ±0.28	232.00 ±2.15	175.67 ±0.52	85.05 ±2.16	175.37± 0.24	146.95 ±0.01	7.97 ±0.05	6.17 ±0.06	9.75± 0.90	10.56 ±0.02	3.70 ±0.10	3.12 ±0.02
May	21.00± 0.07	19.80 ±0.26	210.00± 0.01	215.10± 0.00	85.64 ±7.86	72.25 ±3.16	1.69 ±0.00	1.36 ±0.02	368.00 ±0.26	252.05 ±0.16	184.00± 0.26	115.00 ±0.12	184.00± 0.00	137.05 ±0.04	7.78 ±0.01	8.12 ±0.50	8.76 ±0.20	9.15 ±0.12	3.90 ±0.02	2.15 ±0.02
June	20.25± 0.50	19.00 ±0.20	246.13± 0.00	230.25± 0.21	167.62 ±37.00	242.00 ±52.05	2.53 ±1.13	2.06 ±0.02	1162.93 ±0.08	1321.00 ±0.16	311.86± 1.26	126.15 ±1.25	851.07± 1.17	986.98 ±0.30	7.82 ±0.02	8.32 ±0.16	8.48 ±0.00	8.65 ±0.15	4.80 ±0.00	2.15 ±0.12
July	14.75± 0.27	17.25 ±0.15	312.96± 0.24	311.52± 0.12	356.00 ±127.86	362.00 ±158.25	3.90 ±0.75	1.62 ±0.12	1015.99± 0.19	1215.00 ±0.32	456.00± 21.10	228.02 ±0.62	550.99± 20.91	1194.85 ±1.09	7.68 ±0.01	8.10 ±0.02	7.56 ±0.23	8.15 ±0.05	3.10 ±0.21	3.62 ±0.06
August	21.25± 1.89	18.20 ±0.60	320.10± 0.01	326.02± 0.03	205.05 ±7.84	242.00 ±2.46	2.55 ±0.06	1.52 ±0.03	687.00 ±1.98	782.10 ±0.12	265.00± 3.98	152.00 ±0.16	422.00± 2.00	30.10 ±0.04	8.75 ±0.54	8.56 ±0.05	7.97 ±0.30	8.82 ±0.05	3.30 ±0.04	2.00 ±0.00
September	20.00± 0.50	17.56 ±0.50	256.09± 0.03	260.15± 0.01	116.46 ±22.06	168.62 ±25.00	1.83 ±0.18	1.50 ±0.05	215.90 ±46.00	136.00 ±0.12	154.00± 0.82	220.00 ±2.67	61.90 ±45.18	84.00 ±2.55	7.78 ±0.01	7.62 ±0.52	8.20± 0.17	8.86 ±0.21	3.10 ±0.49	1.25 ±0.25
October	13.50± 0.03	15.25 ±1.21	240.06± 0.09	245.06± 0.00	42.62 ±7.08	62.00 ±2.10	1.64 ±0.73	2.15 ±0.00	159.50 ±12.05	152.00 ±0.50	149.00± 0.08	326.16 ±0.18	10.50 ±11.97	68.00 ±0.35	7.86 ±0.05	7.52 ±0.12	8.88 ±0.22	8.12 ±0.25	1.80 ±0.13	1.16 ±0.01
November	10.00± 0.00	11.50 ±0.05	116.07± 0.00	216.05± 0.05	15.64 ±0.86	15.35 ±2.52	1.87 ±1.18	1.25 ±0.02	144.56 ±72.50	182.00 ±0.15	137.01± 0.30	220.00 ±0.15	7.55 ±72.20	144.15 ±0.03	7.97 ±0.03	7.25 ±0.12	9.52 ±0.11	8.52 ±0.06	2.60 ±0.02	2.25 ±0.02
December	11.25± 0.35	11.65 ±0.10	85.12 ±0.02	90.12± 0.01	42.86 ±7.85	32.00 ±5.12	1.89 ±0.20	1.62 ±2.15	172.65 ±0.14	162.00 ±2.15	159.00± 0.30	165.05 ±0.25	3.65 ±0.16	3.05 ±1.90	6.11 ±0.02	6.85 ±0.06	11.32 ±0.08	10.15 ±0.01	1.57 ±0.12	1.25 ±0.05
Annual Average	**17.38± 0.48**	**15.67 ±0.40**	**180.36± 0.04**	**197.66± 0.06**	**90.44** ±19.01	**104.56 ±21.58**	**1.87 ±0.52**	**1.31 ±0.03**	**438.89 ±11.18**	**372.52 ±0.50**	**203.32± 2.43**	**156.74 ±0.66**	**235.57± 12.87**	**265.65 ±0.55**	**7.66 ±0.07**	**7.46 ±0.17**	**9.61 ±0.24**	**9.57 ±0.08**	**2.89 ±0.11**	**2.50 ±0.05**

Sd±Standard Deviation

Table 4a.6: Monthly variation in Physico-chemical parameter of River Ganga During 2003-2005 at sampling station C

Months	Alkalinity (mg/l)		B.O.D. (mg/l)		C.O.D. (mg/l)		Hardness (mg/l)		Acidity (mg/l)		Chlorides (mg/l)		Calcium (mg/l)		Total Nitrogen (mg/l)		Copper (mg/l)		Magnesium (mg/l)		Iron (mg/l)	
	2003-04	2004-05	2003-04	2004-05	2003-04	2004-05	2003-04	2004-05	2003-04	2004-05	2003-04	2004-05	2003-04	2004-05	2003-04	2004-05	2003-04	2004-05	2003-04	2004-05	2003-04	2004-05
January	104.00± 8.97	65.16± 2.15	1.78 ±0.01	1.85 ±0.26	14.52± 0.54	15.56 ±0.51	55.15 ±0.05	52.50 ±0.05	89.55 ±2.52	60.00± 0.81	5.71 ±6.13	11.05 ±0.02	36.78± 6.17	40.00± 0.02	1.18 ±0.02	2.08 ±0.04	0.04 ±0.02	0.25 ±0.20	89.38 ±29.78	60.82 ±0.15	0.25 ±0.01	0.19± 0.00
February	52.00 ±0.40	84.16 ±2.56	1.17 ±0.00	1.54 ±0.05	11.72± 0.54	12.25 ±0.25	40.25 ±0.05	46.72 ±0.21	75.61 ±1.16	56.35± 0.02	9.53 ±0.09	18.35 ±0.90	34.50± 0.00	32.82± 5.26	2.27 ±0.06	2.52 ±0.50	0.04 ±0.03	0.03 ±0.00	27.98 ±0.24	67.63 ±24.57	0.25 ±0.01	0.16± 0.00
March	97.50 ±0.76	88.62 ±5.16	2.17 ±0.02	2.52 ±0.05	8.53 ±3.13	8.15 ±3.10	50.00 ±0.00	65.10 ±0.10	45.62 ±0.23	42.16± 0.16	12.60 ±0.00	25.16 ±0.08	32.82± 3.01	42.00± 0.05	2.65 ±0.78	2.62 ±1.25	0.04 ±0.01	0.04 ±0.00	83.59 ±14.64	112.39 ±0.24	0.29 ±0.00	0.22± 0.02
April	91.25 ±0.00	86.25 ±.08	2.52 ±0.32	2.52 ±0.68	5.63 ±2.16	6.53 ±0.15	46.78 ±0.18	56.62 ±0.02	43.75 ±0.14	32.25± 0.26	14.39 ±6.71	20.00 ±0.02	27.30± 0.00	25.15± 0.01	2.70 ±0.72	2.42 ±1.32	0.03 ±0.18	0.05 ±0.05	94.78 ±0.88	153.11 ±0.05	0.32 ±0.02	0.25± 0.05
May	82.50 ±3.17	96.32 ±5.68	2.77 ±0.50	2.46 ±0.15	15.50± 0.64	15.26 ±0.42	69.00 ±0.11	62.00 ±0.16	65.62 ±0.45	52.15± 0.16	16.01 ±0.01	25.26 ±0.02	23.68± 0.04	20.00± 0.00	2.16 ±0.15	2.35 ±0.02	0.04 ±0.06	0.06 ±0.00	220.50 ±0.34	204.35 ±0.78	0.41 ±0.03	0.53± 0.05
June	75.35 ±0.13	66.35 ±0.06	2.35 ±0.40	2.52 ±0.16	4.50 ±0.79	3.50 ±0.80	63.00 ±0.41	52.05 ±0.24	45.01 ±0.08	45.25± 0.02	23.59 ±0.36	19.62 ±0.16	21.24± 0.23	22.36± 0.25	2.72 ±0.00	2.82 ±1.62	0.52 ±0.05	0.18 ±0.016	203.18 ±0.88	144.45 ±0.05	0.32 ±0.02	0.53± 0.03
July	87.50 ±3.71	75.25 ±0.18	3.06 ±0.53	2.85 ±0.32	5.67 ±0.49	5.26 ±0.12	67.00 ±0.04	60.02 ±0.12	56.25 ±0.42	52.50± 0.05	18.67 ±0.18	28.00 ±0.00	18.05± 0.53	20.42± 0.15	1.32 ±1.06	1.42 ±0.25	0.06 ±0.02	0.56 ±0.00	238.16 ±2.38	192.67 ±0.15	0.73 ±0.03	0.65± 0.02
August	54.75 ±0.37	82.15 ±0.16	2.73 ±0.45	2.16 ±0.12	10.22± 0.41	8.12 ±1.20	98.00 ±0.20	326.72± 0.16	52.00 ±0.23	62.02± 0.20	14.28 ±0.55	15.85 ±0.32	20.65± 0.58	25.15± 0.12	1.30 ±0.07	1.21 ±0.42	0.06 ±0.00	0.06 ±0.00	376.34 ±1.85	146.72±0. 19	0.72 ±0.02	0.69± 0.00
September	57.78 ±0.19	85.35 ±5.25	3.71 ±0.12	2.85 ±0.16	5.48 ±1.02	5.26 ±0.21	131.00± 0.90	125.00± 0.25	58.03 ±0.16	60.00± 0.16	12.37 ±0.17	18.00 ±0.52	23.20± 0.22	30.16± 0.03	1.56 ±0.06	1.62 ±0.06	0.62 ±0.00	0.05 ±0.03	524.49 ±3.31	461.44 ±1.07	0.63 ±0.01	0.62± 0.06
October	79.71 ±1.00	85.62 ±5.68	3.56 ±0.00	2.37 ±0.06	4.32 ±1.35	4.25 ±1.00	52.60 ±0.05	98.25 ±6.21	34.75 ±0.13	32.50± 0.21	10.05 ±1.06	14.62 ±0.02	25.83± 2.40	25.00± 0.05	0.24 ±0.18	0.15 ±0.00	0.05 ±0.00	0.42 ±0.00	130.25 ±11.43	356.39 ±29.97	0.59 ±0.02	0.56± 0.00
November	95.25 ±7.36	90.35 ±8.61	3.15 ±0.05	3.25 ±0.21	5.07 ±2.05	4.68 ±2.16	74.00 ±1.25	76.00 ±0.83	52.65 ±0.05	60.12± 0.25	9.02 ±2.54	12.35 ±0.03	30.94± 1.12	35.16± 0.16	0.20 ±0.12	0.12 ±0.01	0.05 ±0.00	0.52 ±0.05	209.50 ±0.63	198.70 ±3.26	0.48 ±0.02	0.46± 0.02
December	98.57 ±0.02	85.64 ±0.26	2.06 ±0.02	2.68 ±0.11	12.57 ±0.62	12.51 ±0.22	50.30 ±1.05	72.00 ±0.05	63.71 ±1.39	62.00± 5.25	6.87 ±1.00	12.62 ±0.06	34.59± 0.06	32.15± 0.05	2.46 ±1.06	2.15 ±1.62	0.04 ±0.02	0.01 ±0.01	76.44 ±4.82	193.89 ±0.00	0.22 ±0.20	0.22± 0.00
Annual Average	81.35 ±2.17	82.60 ±12.99	2.59 ±0.20	2.46 ±0.19	8.64 ±1.15	8.44 ±0.85	66.42 ±0.36	91.08 ±0.70	56.88 ±0.58	51.44± 0.63	12.76 ±1.57	18.41 ±0.18	27.47± 1.20	29.20± 0.51	1.73 ±0.36	1.79 ±0.59	0.13 ±0.03	0.19 ±0.04	189.55 ±5.93	220.75 ±2.99	0.43 ±0.03	0.42± 0.02

Sd±Standard Deviation

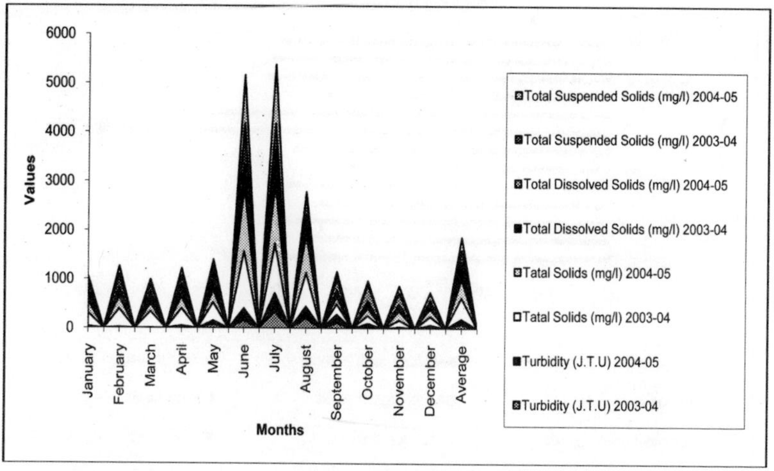

Fig. 4a.14 : Graph showing monthly fluctuation of turbidity, T.S.S., T.S. and T.D.S at Sampling station C

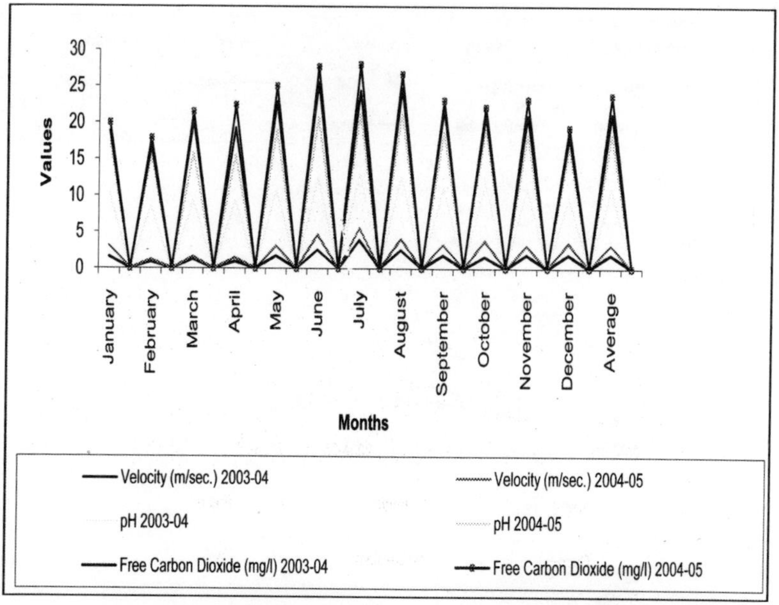

Fig. 4a.15 : Graph showing monthly fluctuation of Velocity, Free CO_2 and pH at Sampling station C

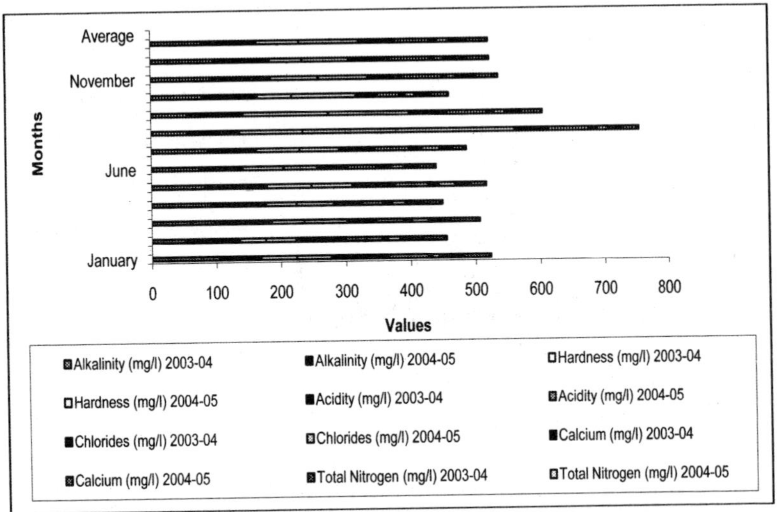

Fig. 4a.16: Graph showing monthly fluctuation of various chemical parameters at Sampling station C

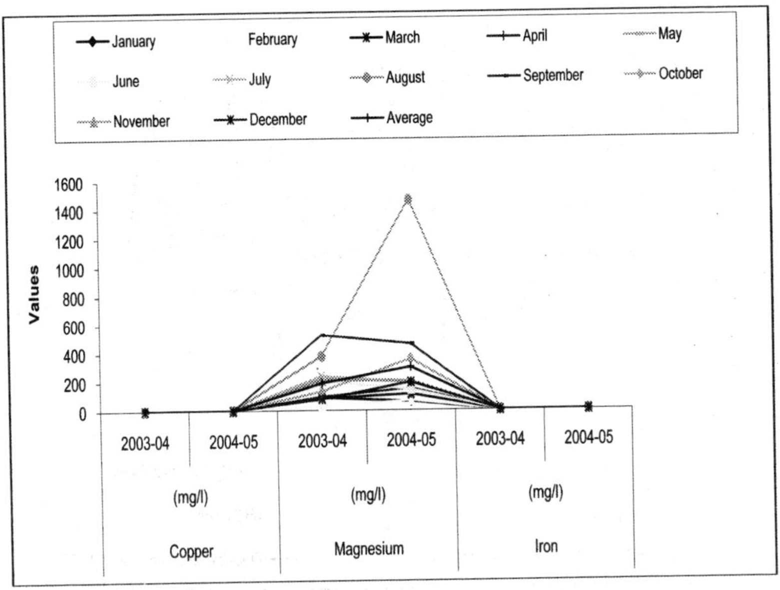

Fig. 4a.17: Graph showing monthly fluctuation of Cu, Mg and Fe at Sampling station C

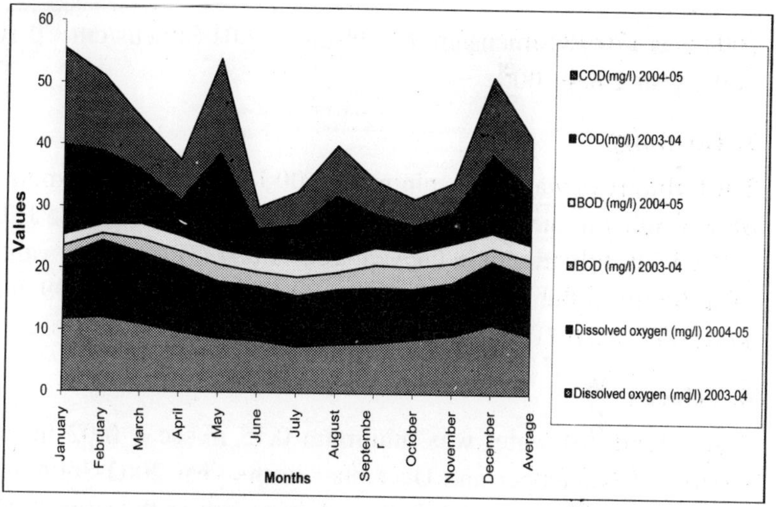

Fig. 4a.18: Graph showing monthly fluctuation of COD, BOD and DO at Sampling station C

Sampling station D

The physico-chemical and heavy metals parameters of sampling station D are presented in Table 4d.1. and Figures (Fig. 4a.19 to Fig. 4a.24).

1. Water Temperature

The water temperature at this sampling spot was minimum 10.15°C ± 0.20 in the December month in the year 2004-2005 and maximum of 24.75°C ± 2.80 in April in the year 2003-2004. The average value in the year 2003-2004 was 17.17°C ± 0.78 and 16.28°C ± 0.40 in the year 2004-2005.

2. Conductivity

The conductivity of river water was observed to be minimum 32.11 Simens/cm ± 0.24 in the August month in the year 2003-2004 and maximum of 360.07 Simens/cm ± 0.00 in the October month in the year 2004-2005. The average values recorded in the year 2003-

2004 was 170.78 Simens/cm ± 0.08 and 202.01 Simens/cm ± 0.10 in the year 2004-2005.

3. Turbidity

The turbidity of water was minimum 0.00 JTU ± 0.00 in the month of November in the year 2003-2004 and maximum of 625.75 JTU ± 105.85 in July month in the year 2003-2004. The average value of turbidity in the year 2003-2004 was 177.90 JTU ± 21.39 and 81.25 JTU ± 16.11 in the year 2004-2005.

4. Velocity

The velocity of water was minimum 0.15 m/sec ± 0.02 in the months of November and December in the year 2003-2004 and maximum of 3.72 m/sec ± 0.06 in July month in the year 2004-2005. The average value of velocity in the year 2003-2004 was 1.09 m/sec ± 0.11 and 1.70 m/sec. ± 0.36 in the year 2004-2005.

5. Total Solids

The total solids were minimum 120.00 mg/l ± 0.06 in February month in the year 2003-2004 and maximum of 1681.00 mg/l ± 00.26 in the July month in the year 2004-2005. The average value of total solids in the year 2003-2004 was 409.72 mg/l ± 9.08 and 442.58 mg/l ± 9.56 in the year 2004-2005.

6. Total Dissolved Solids

The total dissolved solids were minimum 76.51 mg/l ± 0.10 in January month in the year 2003-2004 and maximum of 486.30 mg/l ± 2.54 in the July month in the year 2004-2005. The average value of total solids in the year 2003-2004 was 171.96 mg/l ± 0.48 and 184.97 mg/l ± 8.07 in the year 2004-2005.

7. Total Suspended Solids

The total suspended solids were minimum 7.00 mg/l ± 0.11 in December month in the year 2003-2004 and maximum of 1261.40 mg/l ± 13.88 in the June month in the year 2003-2004. The average value of total solids in the year 2003-2004 was 303.38 mg/l ± 13.95 and 257.58mg/l ± 16.12 in the year 2004-2005.

8. Potential Hydrogenii (pH)

The pH of water were observed to be minimum6.14 ± 0.04 in the December month in the year 2003-2004 and maximum in 8.54± 0.27 in the August month in the year 2004-2005. The average value in the year 2003-2004 was 7.18 ± 0.04 and 7.23 ± 0.30 in the year 2004-2005.

9. Dissolved Oxygen (DO)

The water dissolved oxygen (DO) was minimum 7.15 mg/l ± 0.25 in the July month in the year 2004-2005 and maximum of 12.86 mg/l ± 0.16 in the February month in the year 2004-2005. The average value of DO in the year 2003-2004 was observed to be 9.58 mg/l ± 0.16 and 9.36 mg/l ± 0.33 in the year 2004-2005.

10. Free Carbon Dioxide (CO_2)

The free CO_2 of river water was minimum 1.22 mg/l ± 0.01 in the month of January in the year 2003-2004 and maximum of 4.10 mg/l ± 0.12 in the month of April in the year 2003-2004. The average value of free CO_2 in the year 2003-2004 was 2.53 mg/l ± 0.08 and 2.50 mg/l ± 0.17 in the year 2004-2005.

11. Alkalinity

The alkalinity of river water at this sampling station was minimum 36.00 mg/l ± 2.40 in the month of August in the year 2004-2005 and maximum of 825.00 mg/l ± 0.09 in the month of May in the year 2003-2004. The average value in the year 2003-2005 was 138.53 mg/l ± 5.53 and 75.56 mg/l ± 2.17 in the year 2004-2005.

12. Biochemical Oxygen Demand (BOD)

The water biochemical oxygen demand (BOD) were observed to be minimum 1.15 mg/l ± 0.10 in July in the year 2004-2005 and maximum 5.33 mg/l ± 0.41 in June in the year 2003-2004. The average value in the year 2003-2004 was 3.18 mg/l ± 0.16 and 2.46 mg/l ± 0.27 in the year 2004-2005.

13. Chemical Oxygen Demand (COD)

The chemical oxygen demand (COD) of water was minimum 3.15mg/l ± 2.10 in the month of October in the year 2004-2005 and maximum of 15.58 mg/l ± 0.20 in the month of February in the year 2004-2005. The average value of COD in the year 2003-2004 was 7.89 mg/l ± 1.31 and 8.50 mg/l ± 1.13 in the year 2004-2005.

14. Hardness

The total hardness of water was minimum 32.15 mg/l ± 0.15 in April month in the year 2004-2005 and maximum of 133.00 mg/l ± 0.85 in the month of September in the year 2003-2004. The average value observed in the year 2003-2004 was 69.97 mg/l ± 0.06 and 63.83 mg/l ± 0.80 in the year 2004-2005.

15. Acidity

The acidity of water was observed minimum of 18.25 mg/l ± 0.09 in July month in the year 2003-2005 and maximum of 82.15 mg/l ± 2.15 in the month of January in the year 2004-2005.The average value observed in the year 2003-2004 was 49.33 mg/l ± 0.08 and 57.58 mg/l ± 0.84 in the year 2004-2005.

16. Chlorides

The Chlorides in water was minimum 6.20 mg/l ± 3.10 in the month of January in the year 2004-2005 and maximum of 26.02 mg/l ± 0.81 in the month of July in the year 2004-2005.The average value of Chlorides in the year 2003-2004 was 15.40 mg/l ± 0.15and 12.74 mg/l ± 0.39 in the year 2004-2005.

17. Calcium

The calcium in riverwater at sampling station 'D ' was minimum 10.21 mg/l ± 0.03 in month of December in the year 2003-2004 and maximum of 39.28 mg/l ± 0.27 in January in the year 2003-2004. The average value in the year 2003-2004 was 15.40 mg/l ± 0.15 and 22.53 mg/l ± 0.89 in the year 2004-2005.

18. Total Nitrogen

The water total nitrogen in water at this sampling station 'D' was minimum 0.19 mg/l ± 0.02 in November month in the year 2004-2005 and maximum of 2.87 mg/l ± 0.32 in the month of April in the year 2003-2004. The average value in the year 2003-2004 was 1.84 mg/l ± 0.56 and 1.71 mg/l ± 0.22 in the year 2004-2005.

19. Copper

The copper in riverwater at sampling station 'D' was minimum 0.02 mg/l ± 0.00 in the months of January, April, September, October, December in the year 2003-2004 and maximum of 0.62 mg/l ± 0.03 in September in the year 2004-2005. The average value in the year 2003-2004 was 0.09 mg/l ± 0.00 and 0.21 mg/l ± 0.04 in the year 2004-2005.

20. Magnesium

The Magnesium in riverwater at sampling station 'D' was minimum 63.25 mg/l ± 0.54 in month of April in the year 2004-2005 and maximum of 731.84 mg/l ± 2.19 in September in the year 2003-2004. The average value in the year 2003-2004 was 275.58 mg/l ± 1.60 and 200.92 mg/l ± 5.07 in the year 2004-2005.

21. Iron

The concentration of iron in riverwater at sampling station 'D' was observed minimum of 0.16 mg/l ± 0.00 in January month in the year 2003-2004 and maximum of 0.68 mg/l ± 0.00 in the month of July in the year 2003-2004. The average value observed in the year 2003-2004 was 00.45 mg/l ± 0.05 and 0.35 mg/l ± 0.03 in the year 2004-2005.

Table 4a.7: Monthly variation in Physico-chemical parameter of River Ganga During 2003-2005 at sampling station D

Months	Temp. (°C)		Conductivity (Simens/cm.)		Turbidity (J.T.U.)		Velocity (m/sec.)		Total Solids (mg/l)		Total Dissolved Solids (mg/l)		Total Suspended Solids (mg/l)		pH		Dissolved Oxygen (mg/l)		Free Carbon Dioxide (mg/l)	
	2003-04	2004-05	2003-04	2004-05	2003-04	2004-05	2003-04	2004-05	2003-04	2004-05	2003-04	2004-05	2003-04	2004-05	2003-04	2004-05	2003-04	2004-05	2003-04	2004-05
January	11.15± 1.07	10.65± 0.15	80.13± 0.01	82.12± 0.00	28.00± 1.08	15.62± 0.15	0.20± 0.01	1.25± 0.02	165.00± 0.24	256.00± 2.15	76.51± 0.10	187.08± 3.50	162.49± 0.05	68.92± 1.35	7.35± 0.10	6.96± 2.14	11.10± 0.00	11.15± 0.06	1.22± 0.01	1.25± 0.12
February	17.00± 1.07	12.15± 0.02	70.21± 0.03	65.35± 0.16	26.78± 2.10	9.61± 0.08	0.25± 0.05	0.72± 1.50	120.00± 0.06	325.00± 12.50	102.00± 0.20	95.65± 2.46	265.46± 29.43	229.35± 10.04	7.55± 0.05	6.62± 0.16	12.56± 0.16	12.86± 0.16	1.75± 0.05	1.46± 0.25
March	17.00± 0.80	15.16± 1.00	16016± 0.05	50.17± 0.10	40.00± 3.56	2.80± 0.12	0.35± 0.00	0.25± 0.20	329.60± 0.28	252.16± 0.16	89.20± 0.34	106.20± 2.46	240.40± 0.06	145.60± 2.30	6.18± 0.00	6.15± 0.00	11.25± 0.08	12.32± 0.01	1.85± 0.06	3.50± 0.00
April	24.75± 2.80	19.15± 0.16	185.18± 0.02	120.18± 0.10	45.66± 6.85	13.15± 5.64	0.45± 0.03	0.56± 0.16	300.00± 0.24	360.00± 0.10	95.15± 0.20	152.25± 20.00	204.85± 25.12	207.75± 19.90	6.15± 0.00	7.15± 0.12	9.08± **0.00**	10.16± 2.20	4.10± 0.12	3.20± 0.02
May	22.00± 0.40	19.75± 1.54	210.11± 0.08	214.16± 0.00	162.56± 37.24	72.15± 2.46	0.92± 0.02	2.20± 0.20	293.60± 0.79	378.82± 0.86	115.05± 0.46	163.08± 6.20	177.95± 19.41	215.74± 5.34	7.13± 0.00	7.25± 0.06	9.11± 0.22	7.62± 0.02	3.17± 0.04	3.60± 0.01
June	20.50± 0.57	19.85± 1.26	260.12± 0.00	310.15± 0.05	275.00± 32.14	240.00± 32.08	0.97± 0.18	3.15± 1.68	1435.00± 0.18	568.42± 0.72	173.60± 0.03	262.08± 5.26	1261.40± 13.88	306.34± 4.54	7.30± 0.03	7.85± 0.25	8.03± 0.12	8.62± 0.03	3.52± 0.40	3.80± 0.00
July	19.75± 0.95	17.15± 1.72	310.96± 0.07	312.35± 0.16	625.75± 105.85	254.00± 132.80	2.41± 0.06	3.72± 0.06	1125.00± 73.60	1681.00± 0.26	198.00± 1.74	486.30± 2.54	927.00± 71.86	1194.70± 2.28	7.20± 0.02	7.16± 0.00	7.67± 0.05	7.15± 0.25	3.93± 0.13	4.20± 0.02
August	18.00± 0.00	17.25± 1.25	32.11± 0.24	250.84± 0.20	500.05± 37.06	115.65± 2.15	2.47± 0.06	2.75± 0.15	539.00± 0.05	782.00± 68.00	174.02± 0.08	162.00± 1.64	54.02± 0.02	620.00± 66.36	8.50± 0.03	8.54± 0.27	8.11± 0.35	7.36± 0.50	3.13± 0.04	3.50± 0.02
September	17.50± 1.91	19.15± 0.00	215.10± 0.37	350.50± 0.16	356.06± 22.75	106.00± 6.85	2.35± 0.05	1.25± 0.23	367.46± 29.63	192.00± 24.00	278.00± 0.30	162.80± 0.16	113.00± 0.31	29.20± 23.84	7.55± 0.06	7.62± 0.21	8.55± 0.71	8.16± 0.12	2.56± 0.04	2.50± 0.21
October	19.25± 0.50	19.85± 0.25	230.07± 0.03	360.07± 0.00	70.00± 8.06	62.00± 6.10	2.37± 0.82	1.15± 0.00	197.00± 3.68	165.00± 5.65	295.00± 1.72	152.00± 0.30	124.99± 6.62	13.00± 5.35	7.17± 0.06	7.25± 0.06	8.65± 0.11	8.25± 0.20	1.30± 0.00	1.50± 0.16
November	14.25± 0.04	15.15± 0.06	210.12± 0.01	218.06± 0.02	0.00± 0.00	52.00± 6.10	0.15± 0.02	1.82± 0.06	170.00± 3.68	152.00± 0.12	299.00± 0.61	106.20± 52.12	102.00± 0.54	45.80± 52.00	7.97± 0.09	7.35± 0.01	9.68± 0.13	9.24± 0.16	2.45± 0.00	2.50± 1.23
December	10.37± 0.00	10.15± 0.20	85.14± 0.00	90.15± 0.24	8.00± 0.00	32.05± 2.46	0.15± 0.02	1.62± 0.10	175.00± 0.14	198.54± 0.16	168.00± 0.03	184.00± 0.25	7.00± 0.11	14.54± 0.09	6.14± 0.04	6.84± 0.26	11.15± 0.01	10.24± 0.24	1.43± 0.09	1.35± 0.00
Annual Average	17.17± 0.78	16.28± 0.63	170.78± 0.08	202.01± 0.10	177.90± 21.39	81.25± 16.11	1.09± 0.11	1.70± 0.36	409.72± 9.08	442.58± 9.56	171.96± 0.48	184.97± 8.07	303.38± 13.95	257.58± 16.12	7.18± 0.04	7.23± 0.30	9.58± 0.16	9.36± 0.33	2.53± 0.08	2.50± 0.17

Sd±Standard Deviation

Table 4a.8: Monthly variation in Physico-chemical parameter of River Ganga During 2003-2005 at sampling station D

Months	Alkalinity		B.O.D.		C.O.D.		Hardness		Acidity		Chlorides		Calcium		Total Nitrogen		Copper		Magnesium		Iron	
	2003-04	2004-05	2003-04	2004-05	2003-04	2004-05	2003-04	2004-05	2003-04	2004-05	2003-04	2004-05	2003-04	2004-05	2003-04	2004-05	2003-04	2004-05	2003-04	2004-05	2003-04	2004-05
January	64.75± 18.65	92.50± 0.12	2.19± 0.08	3.16± 0.00	14.17± 2.16	12.15± 0.16	60.00± 0.06	65.20± 0.06	65.67± 0.06	82.15± 2.15	9.76± 0.05	6.20± 3.10	39.28± 0.27	32.15± 5.12	2.00± 0.00	2.15± 0.01	0.02± 0.00	0.02± 0.02	136.31± 1.02	160.80± 24.62	0.16± 0.00	0.20± 0.00
February	67.40± 29.65	80.62± 2.15	2.20± 0.03	3.15± 2.15	11.62± 0.40	15.60± 0.20	61.,00± 0.00	50.15± 2.10	55.00± 0.00	72.15± 2.16	11.52± 0.08	8.25± 0.25	35.42± 0.16	25.08± 0.00	2.18± 1.02	2.42± 0.05	0.03± 0.00	0.03± 0.05	117.49± 0.58	121.98± 10.22	0.22± 0.12	0.26± 0.05
March	87.50± 3.15	98.82± 6.56	2.32± 0.21	2.32± 0.20	8.56± 3.10	8.25± 2.08	45.80± 0.04	50.00± 0.05	46.80± 0.02	52.12± 0.26	15.00± 0.05	11.25± 0.00	32.82± 1.39	20.00± 0.00	2.42± 1.64	2.50± 0.05	0.03± 0.00	0.05± 0.01	162.71± 6.52	145.96± 0.24	0.24± 0.00	0.28± 0.08
April	112.50±3 .17	97.16± 0.62	2.62± 0.03	2.62± 0.10	6.54± 3.13	5.20± 2.10	53.24± 0.04	32.15± 0.15	45.00± 0.20	52.45± 0.16	18.15± 0.15	12.52± 0.02	27.56± 0.05	19.15± 0.05	2.87± 0.32	2.56± 0.06	0.02± 0.00	0.02± 0.16	**220.47±** 0.24	63.25± 0.54	0.46± 0.14	0.35± 0.05
May	825.00±0 .09	87.52± 2.46	3.63± 0.09	2.52± 0.20	14.60± 0.45	11.00± 0.15	57.50± 0.54	52.00± 0.26	64.37± 0.13	72.15± 0.86	19.17± 0.13	15.21± 0.00	25.88± 0.53	18.52± 0.02	2.14± 0.01	2.10± 0.01	0.35± 0.00	0.03± 0.01	218.36± 3.36	162.89± 1.16	0.48± 0.00	0.50± 0.02
June	75.25± 0.12	65.12± 0.16	5.33± 0.41	2.32± 0.10	4.70± 0.32	6.50± 0.15	57.50± 0.05	62.00± 0.21	40.00± 0.08	52.50± 0.28	22.72± 0.42	20.00± 0.10	21.24± 0.23	20.12± 0.12	2.37± 1.84	2.32± 0.03	0.30± 0.00	0.52± 0.05	235.77± 1.51	203.76± 0.44	0.60± 0.01	0.32± 0.01
July	98.17± 7.08	52.00± 2.46	4.79± 0.45	1.15± 0.10	5.21± 0.46	5.12± 0.12	70.15± 0.05	65.00± 0.02	18.25± 0.09	56.72± 0.84	19.37± 0.08	26.02± 0.81	19.37± 0.50	19.12± 0.21	1.35± 0.05	1.25± 1.02	0.05± 0.00	0.03± 0.00	334.80± 2.19	223.23± 0.92	0.68± 0.00	0.53± 0.02
August	68.25± 0.05	36.00± 2.40	4.40± 0.36	1.68± 0.00	3.20± 0.42	10.62± 0.21	98.00± 0.24	65.15± 0.10	65.00± 0.18	52.00± 0.20	18.31± 0.51	15.25± 0.15	18.31± 0.51	20.52± 0.60	1.20± 0.02	1.20± 0.00	0.04± 0.00	0.05± 0.33	513.93± 0.15	314.45± 2.43	0.57± 0.10	0.61± 0.06
September	77.50± 0.24	64.06± 0.13	3.60± 0.11	2.32± 0.20	4.13± 0.44	6.10± 2.10	133.00± 0.85	98.00± 0.08	45.50± 0.12	54.00± 0.82	15.16± 0.17	11.25± 0.15	15.16± 0.17	23.15± 0.12	1.75± 0.05	1.35± 0.26	0.02± 0.00	0.62± 0.03	731.84± 2.19	364.18± 0.19	0.56± 0.00	0.51± 0.05
October	84.51± 3.15	62.15± 1.25	2.75± 0.05	3.08± 0.00	5.12± 2.01	3.15± 2.10	59.00± 0.09	62.00± 0.06	34.00± 0.00	32.25± 0.12	14.20± 0.05	10.20± 0.00	14.20± 0.05	25.15± 2.12	1.36± 0.00	0.21± 0.12	0.02± 0.00	0.62± 0.02	117.16± 1.07	179.29± 10.02	0.51± 0.00	0.25± 0.00
November	47.56± 0.40	85.15± 5.62	2.08± 0.00	3.00± 0.00	4.56± 2.54	6.80± 2.05	78.00± 0.00	82.00± 2.00	52.85± 0.05	50.26± 0.08	11.26± 0.04	8.52± 0.05	11.26± 0.04	22.15± 2.16	0.24± 1.00	0.19± 0.02	0.19± 0.00	0.52± 0.05	301.50± 0.24	291.19± 0.78	0.44± 0.15	0.20± 0.00
December	54.00± 0.14	85.68± 2.16	2.17± 0.03	2.15± 0.10	12.30± 0.24	11.52± 2.10	67.00± 0.06	62.25± 2.10	59.71± 0.09	62.25± 2.13	10.21± 0.03	8.25± 0.02	10.21± 0.03	25.25± 0.20	2.15± 0.80	2.25± 1.05	0.02± 0.00	0.03± 0.10	216.32± 0.15	180.02± 9.24	0.46± 0.02	0.19± 0.02
Annual Average	138.53±5 .53	**75.56±** 2.17	3.18± 0.16	2.46± 0.27	7.89± 1.31	8.50± 1.13	69.97± 0.06	63.83± 0.60	49.33± 0.08	57.58± 0.84	15.40± 0.15	12.74± 0.39	15.40± 0.15	22.53± 0.89	1.84± 0.56	1.71± 0.22	0.09± 0.00	0.21± 0.04	275.58± 1.60	200.92± 5.07	0.45± 0.05	0.35± 0.03

Sd±Standard Deviation

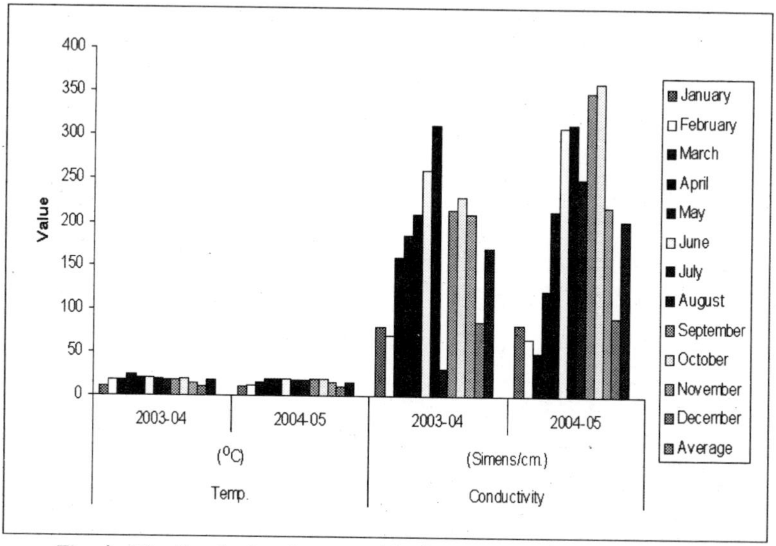

Fig. 4a.19 : Graph showing monthly fluctuation of temperature and conductivity at sampling station D

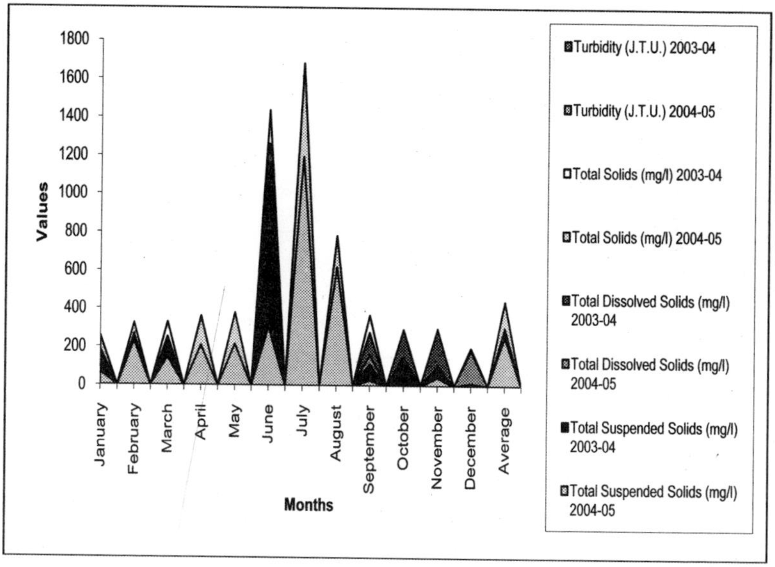

Fig. 4a.20 : Graph showing monthly fluctuation of turbidity, T.S.S., T.S. and T.D.S at Sampling station D

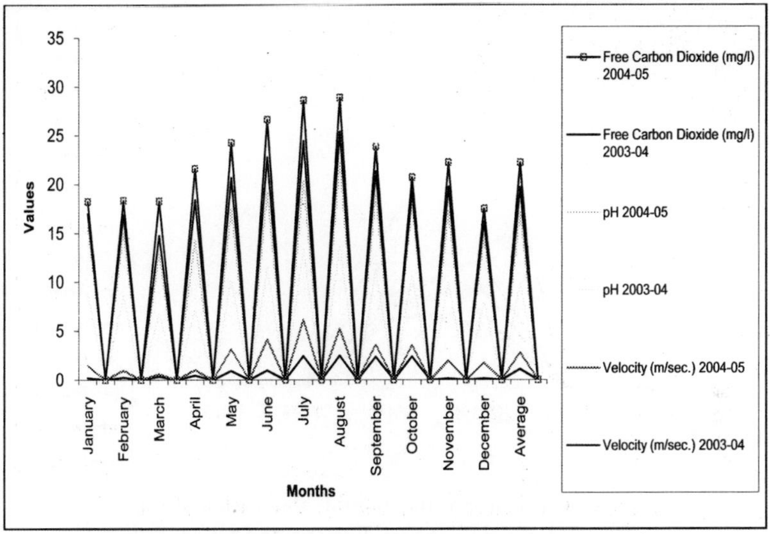

Fig. 4a.21 : Graph showing monthly fluctuation of Velocity, Free CO_2 and pH at Sampling station D

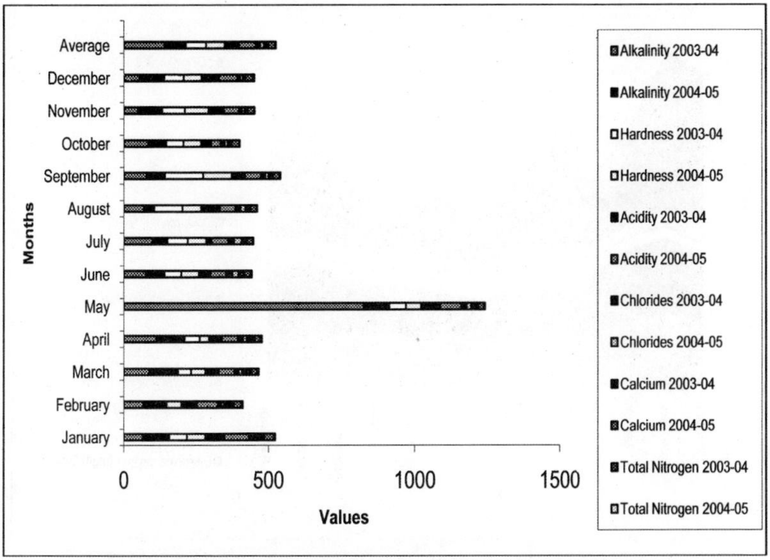

Fig. 4a.22: Graph showing monthly fluctuation of various chemical parameters at Sampling station D

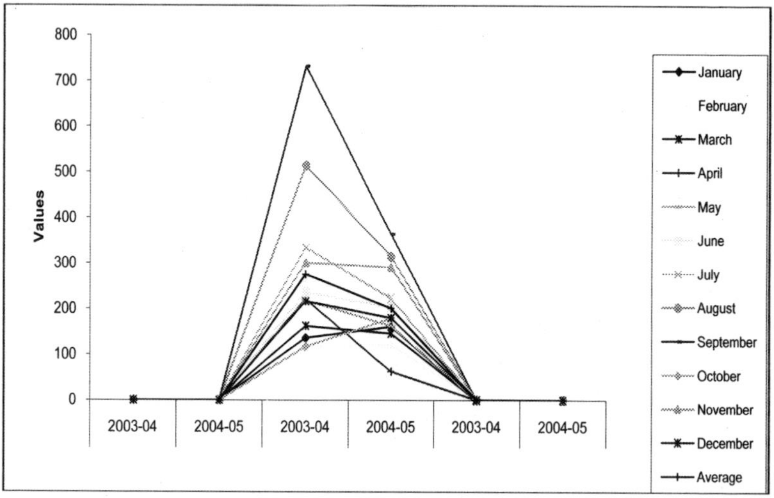

Fig. 4a.23: Graph showing monthly fluctuation of Cu, Mg and Fe at Sampling station D

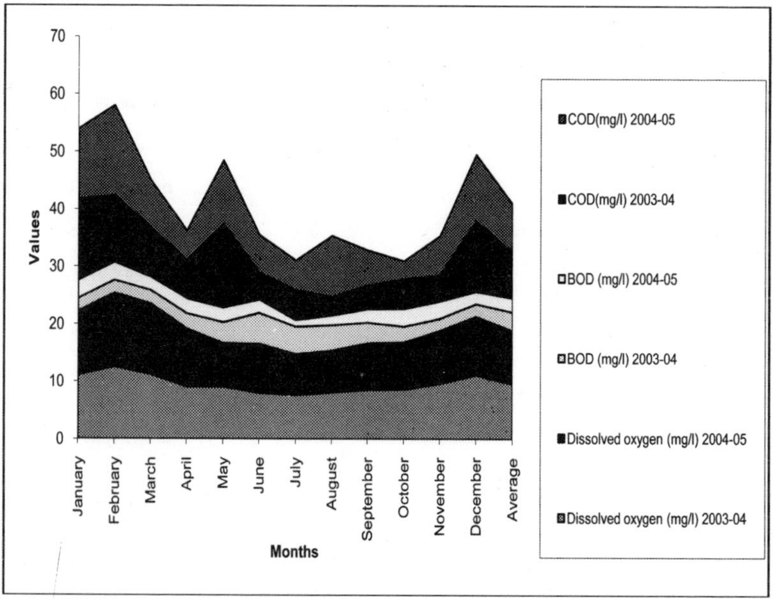

Fig. 4a.24: Graph showing monthly fluctuation of COD, BOD and DO at Sampling station D

Sampling station E

The physico-chemical and heavy metals parameters of sampling station E are presented in Table 4e.1 and Figures (Fig.4a.25 to Fig. 4a.30).

1. Water Temperature

The water temperature at this sampling spot was minimum 10.25°C ± 0.25 in the November month in the year 2003-2004 and maximum of 22.50°C ± 0.01 in March in the year 2003-2004. The average value in the year 2003-2004 was 17.06°C ± 0.33 and 15.57°C ± 0.08 in the year 2004-2005.

2. Conductivity

The conductivity of river water was observed to be minimum 50.15 Simens/cm ± 00.05 in the March month in the year 2004-2005 and maximum of 335.11 Simens/cm ± 0.21 in the August month in the year 2004-2005. The average values recorded in the year 2003-2004 was 180.28 Simens/cm ± 0.08 and 181.05 Simens/cm ± 0.04 in the year 2004-2005.

3. Turbidity

The turbidity of water was minimum 0.00 JTU ± 0.00 in the month of October in the year 2004-2005 and maximum of 635.66 JTU ± 90.62 in July month in the year 2004-2005. The average value of turbidity in the year 2003-2004 was 125.59 JTU ± 22.21 and 161.09 JTU ± 25.76 in the year 2004-2005.

4. Velocity

The velocity of water was minimum 0.12 m/sec ± 0.00 in the months of October in the year 2004-2005 and maximum of 3.15 m/sec ± 0.16 in July month in the year 2004-2005. The average value of turbidity in the year 2003-2004 was 0.84 m/sec ± 0.13 and 1.17 m/sec. ± 0.10 in the year 2004-2005.

5. Total Solids

The total solids were minimum 1.50 mg/l ± 0.62 in February month in the year 2003-2004 and maximum of 1216.00 mg/l ± 00.12 in the July month in the year 2004-2005. The average value of total solids in the year 2003-2004 was 334.61 mg/l ± 3.00 and 360.76 mg/l ± 0.95 in the year 2004-2005.

6. Total Dissolved Solids

The total dissolved solids were minimum 82.56 mg/l ± 0.17 in March month in the year 2004-2005 and maximum of 286.00 mg/l ± 0.25 in the September month in the year 2004-2005. The average value of total solids in the year 2003-2004 was 157.73 mg/l ± 5.76 and 164.05 mg/l ± 0.77 in the year 2004-2005.

7. Total Suspended Solids

The total suspended solids were minimum 3.39 mg/l ± 13.80 in December month in the year 2003-2004 and maximum of 1053.00 mg/l ± 0.50 in the July month in the year 2004-2005. The average value of total solids in the year 2003-2004 was 176.88 mg/l ± 8.13 and 247.38 mg/l ± 0.51 in the year 2004-2005.

8. Potential Hydrogenii (pH)

The pH of water were observed to be minimum 5.94 ± 0.02 in the February month in the year 2003-2004 and maximum in 8.90± 0.07 in the November month in the year 2003-2004. The average value in the year 2003-2004 was 7.49 ± 0.11 and 7.18 ± 0.06 in the year 2004-2005.

9. Dissolved Oxygen (DO)

The water dissolved oxygen (DO) was minimum 6.17 mg/l ± 0.12 in the June month in the year 2003-2004 and maximum of 14.35 mg/l ± 0.00 in the December month in the year 2003-2004. The average value of DO in the year 2003-2004 was observed to be 9.24 mg/l ± 0.33 and 9.71 mg/l ± 0.08 in the year 2004-2005.

10. Free Carbon Dioxide (CO_2)

The free CO_2 of river water was minimum 1.13 mg/l ± 0.07 in the month of January in the year 2003-2004 and maximum of 4.12 mg/l ± 0.12 in the month of July in the year 2004-2005. The average value of free CO_2 in the year 2003-2004 was 2.33 mg/l ± 0.05 and 2.48 mg/l ± 0.05 in the year 2004-2005.

11. Alkalinity

The alkalinity of river water at this sampling station was minimum 17.25 mg/l ± 0.05 in the month of November in the year 2003-2004 and maximum of 87.50 mg/l ± 0.51 in the month of March in the year 2003-2004. The average value in the year 2003-2004 was 60.14 mg/l ± 0.64 and 70.03 mg/l ± 1.40 in the year 2004-2005.

12. Biochemical Oxygen Demand (BOD)

The water biochemical oxygen demand (BOD) was observed to be minimum 1.65 mg/l ± 0.20 in February in the year 2004-2005 and maximum 5.68 mg/l ± 00.74 in June in the year 2003-2004. The average value in the year 2003-2004 was 3.33 mg/l ± 0.15 and 22.37 mg/l ± 0.14 in the year 2004-2005.

13. Chemical Oxygen Demand (COD)

The chemical oxygen demand (COD) of water was minimum 3.46mg/l ± 0.03 in the month of November in the year 2004-2005 and maximum of 16.24 mg/l ± 0.06 in the month of January in the year 2004-2005. The average value of COD in the year 2003-2004 was 8.95 mg/l ± 0.84 and 8.82 mg/l ± 0.47 in the year 2004-2005.

14. Hardness

The total hardness of water was minimum 32.00 mg/l ± 0.02 in March month in the year 2004-2005 and maximum of 162.00 mg/l ± 0.34 in the month of August in the year 2004-2005. The average value observed in the year 2003-2004 was 68.06 mg/l ± 0.25 and 74.19 mg/l ± 0.11 in the year 2004-2005.

15. Acidity

The acidity of water was observed minimum of 25.50 mg/l ± 0.60 in July month in the year 2004-2005 and maximum of 71.25 mg/l ± 0.26 in the month of August in the year 2003-2004.The average value observed in the year 2003-2004 was 54.69 mg/l ± 0.52 and 46.12 mg/l ± 0.10 in the year 2004-2005.

16. Chlorides

The Chlorides in water was minimum 7.71 mg/l ± 0.06 in the month of January in the year 2003-2004 and maximum of 26.08 mg/l ± 0.02 in the month of July in the year 2004-2005.The average value of Chlorides in the year 2003-2004 was 13.72 mg/l ± 0.28 and 19.87 mg/l ± 0.10 in the year 2004-2005.

17. Calcium

The calcium in river water at sampling station 'E ' was minimum 15.06 mg/l ± 0.73 in month of May in the year 2003-2004 and maximum of 48.00 mg/l ± 0.01 in January in the year 2004-2005. The average value in the year 2003-2004 was 23.96 mg/l ± 0.26 and 28.42 mg/l ± 0.09 in the year 2004-2005.

18. Total Nitrogen

The water total nitrogen in water at this sampling station 'E' was minimum 0.10 mg/l ± 0.00 in November month in the year 2004-2005 and maximum of 2.66 mg/l ± 0.81 in the month of April in the year 2004-2005. The average value in the year 2003-2004 was 1.93 mg/l ± 0.45 and 1.89 mg/l ± 0.34 in the year 2004-2005.

19. Copper

The copper in river water at sampling station 'E' was minimum 0.02 mg/l ± 0.00 in the months of February, March in the year 2003-2004 and November in the year 2004-2005. The maximum value copper was observed 0.67 mg/l ± 0.00 in the month of September in the year 2004-2005. The average value in the year

2003-2004 was 0.16 mg/l ± 0.00 and 0.10 mg/l ± 0.04 in the year 2004-2005.

20. Magnesium

The magnesium in river water at sampling station 'E' was minimum 33.77 mg/l ± 0.15 in month of March in the year 2004-2005 and maximum of 705.24 mg/l ± 1.17 in September in the year 2004-2005.The average value in the year 2003-2004 was 214.56 mg/l ± 1.67 and 222.66 mg/l ± 0.41 in the year 2004-2005.

21. Iron

The concentration of iron in river water at sampling station 'E' was observed minimum of 0.17 mg/l ± 0.00 in January month in the year 2003-2004 and maximum of 0.76 mg/l ± 0.05 in the month of July in the year 2003-2004. The average value observed in the year 2003-2004 was 0.44 mg/l ± 0.05 and 0.47 mg/l ± 0.06 in the year 2004-2005.

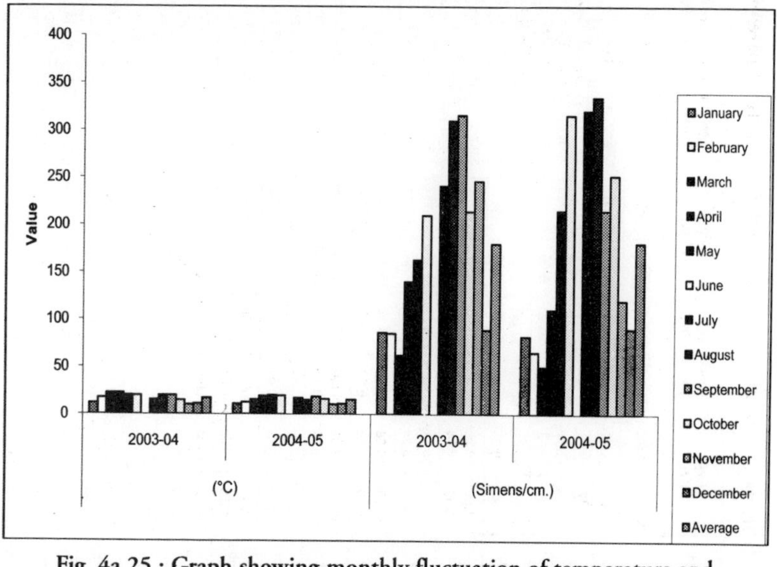

Fig. 4a.25 : Graph showing monthly fluctuation of temperature and conductivity at sampling station E

Table 4a.9: Monthly variation in Physico-chemical parameter of River Ganga During 2003-2005 at sampling station E

Months	Temp. (°C)		Conductivity (Simens/cm.)		Turbidity (J.T.U.)		Velocity (m/sec.)		Total Solids (mg/l)		Total Dissolved Solids (mg/l)		Total Suspended Solids (mg/l)		pH		Dissolved Oxygen (mg/l)		Free Carbon Dioxide (mg/l)	
	2003-04	2004-05	2003-04	2004-05	2003-04	2004-05	2003-04	2004-05	2003-04	2004-05	2003-04	2004-05	2003-04	2004-05	2003-04	2004-05	2003-04	2004-05	2003-04	2004-05
January	11.75± 0.35	11.00± 0.01	86.21± 0.23	82.15± 0.05	15.65± 1.08	46.00± 1.25	0.45± 0.00	0.32± 0.02	158.00± 0.26	232.00± 0.01	194.00± 0.13	85.15± 0.16	12.90± 6.62	146.84± 0.15	7.29± 0.03	7.18± 0.15	12.10± 0.08	11.20± 0.00	1.13± 0.07	1.20± 0.00
February	17.25± 0.35	12.56± 0.02	85.17± 0.03	65.26± 0.02	1.00± 0.00	28.00± 0.21	0.35± 0.03	0.56± 0.05	1.50± 0.62	268.04± 2.65	86.20± 0.02	164.00± 2.16	28.99± 0.31	104.04± 0.49	5.94± 0.02	7.36± 0.02	11.26± 0.09	12.06± 0.01	1.62± 0.13	1.62± 0.06
March	22.50± 0.01	15.62± 0.06	62.18± 0.16	50.15± 0.05	5.00± 0.05	15.06± 0.00	0.25± 0.00	0.57± 0.05	394.00± 2.25	326.78± 0.16	104.78± 0.34	82.56± 0.17	90.92± 0.33	244.22± 0.01	7.35± 0.00	6.25± 0.00	10.60± 0.70	11.15± 0.02	1.50± 0.14	1.82± 0.03
April	22.50± 0.00	19.20± 0.08	140.21± 0.19	110.18± 0.01	32.57± 6.42	19.25± 2.11	0.50± 0.00	0.82± 0.08	371.28± 0.10	254.26± 1.06	276.71± 25.22	98.25± 0.65	94.57± 25.12	156.01± 0.41	7.62± 0.02	6.15± 0.00	9.90± 0.85	9.01± 0.01	2.70± 0.06	2.14± 0.10
May	20.00± 0.57	20.15± 0.24	163.10± 0.01	215.15± 0.02	87.67± 7.84	98.25± 82.00	1.12± 0.01	0.98± 0.25	370.04± 0.79	242.00± 2.14	176.67± 20.20	126.00± 0.52	191.37± 19.41	116.00± 1.62	7.32± 0.01	7.36± 0.02	8.18± 0.19	9.80± 0.15	2.60± 0.00	3.85± 0.08
June	19.75± 0.50	19.68± 0.16	210.12± 0.02	316.18± 0.00	240.00± 32.46	182.00± 62.00	1.32± 0.04	2.16± 0.02	315.67± 0.74	1026.00± 2.68	149.61± 14.14	162.60± 0.02	19.24± 0.07	144.00± 0.86	7.56± 0.05	7.42± 0.05	6.17± 0.12	8.62± 0.08	4.10± 0.00	3.65± 0.06
July	15.25± 0.35	17.06± 0.04	240.96± 0.27	320.92± 0.02	467.00± 172.04	635.66± 90.62	1.00± 0.12	3.15± 0.16	1137.79± 2.36	1216.00± 0.12	238.00± 0.12	182.00± 1.82	899.79± 2.24	1053.00± 0.50	7.19± 0.15	7.12± 0.00	5.20± 0.23	7.65± 0.05	2.70± 0.08	4.12± 0.12
August	19.75± 0.38	15.06± 0.05	310.11± 0.00	335.11± 0.21	315.00± 2.40	468.25± 27.36	1.10± 0.00	2.16± 0.02	878.71± 26.50	726.00± 0.05	135.26± 0.15	164.06± 0.81	78.00± 0.62	38.06± 0.76	8.52± 0.58	8.68± 0.25	7.24± 0.80	8.16± 0.18	3.50± 0.04	3.15± 0.05
September	19.75± 0.50	18.62± 0.02	316.08± 0.02	215.09± 0.06	217.00± 32.62	256.08± 32.008	1.40± 0.11	2.42± 0.41	163.00± 0.19	162.00± 0.21	134.01± 0.50	286.00± 0.25	289.22± 1.91	72.00± 0.85	7.50± 0.03	7.62± 0.08	8.20± 0.71	8.62± 0.25	2.60± 0.00	2.52± 0.06
October	14.75± 0.35	16.06± 0.05	214.01± 0.03	252.06± 0.03	56.00± 6.00	0.56± 0.26	1.61± 1.06	0.12± 0.00	178.50± 1.70	132.00± 2.06	165.60± 8.32	212.00± 1.42	292.81± 28.38	80.00± 0.64	7.35± 0.15	7.10± 0.07	10.16± 0.15	8.75± 0.12	1.50± 0.02	2.10± 0.00
November	10.25± 0.25	10.52± 0.01	246.03± 0.00	120.15± 0.05	25.00± 0.00	52.00± 8.15	0.55± 0.09	0.15± 0.02	189.71± 0.19	182.00± 0.06	94.71± 0.02	234.00± 1.06	121.67± 0.61	104.00± 0.19	8.90± 0.07	7.62± 0.06	9.70± 0.06	9.25± 0.06	2.20± 0.01	2.35± 0.03
December	11.25± 0.35	11.25± 0.20	89.12± 0.02	90.16± 0.00	45.20± 5.62	52.00± 3.16	0.45± 0.13	0.16± 0.02	185.63± 0.35	162.00± 0.16	170.47± 0.12	172.00± 0.21	3.39± 13.80	10.00± 0.05	7.37± 0.15	6.26± 0.02	14.35± 0.00	0.25± 0.00	1.83± 0.06	1.25± 0.02
Annual Average	17.06± 0.33	15.57± 0.08	180.28± 0.08	181.05± 0.04	125.59± 22.21	161.09± 25.76	0.84± 0.13	1.17± 0.10	334.61± 3.00	360.76± 0.95	157.73± 5.76	164.05± 0.77	176.88± 8.13	247.38± 0.51	7.49± 0.11	7.18± 0.06	9.42± 0.33	9.71± 0.08	2.33± 0.05	2.48± 0.05

Sd±Standard Deviation

Table 4a.10: Monthly variation in Physico-chemical parameter of River Ganga During 2003-2005 at sampling station E

Months	Alkalinity		B.O.D.		C.O.D.		Hardness		Acidity		Chlorides		Calcium		Total Nitrogen		Copper		Magnesium		Iron	
	2003-04	2004-05	2003-04	2004-05	2003-04	2004-05	2003-04	2004-05	2003-04	2004-05	2003-04	2004-05	2003-04	2004-05	2003-04	2004-05	2003-04	2004-05	2003-04	2004-05	2003-04	2004-05
January	25.25± 0.17	75.84± 2.16	3.50± 0.00	2.08± 0.02	12.54± 0.25	16.24± 0.06	58.50± 0.04	68.25± 0.25	60.00± 0.00	52.16± 0.06	7.71± 0.06	12.25± 0.01	36.20± 0.03	48.00± 0.01	2.37± 0.06	2.32± 0.02	0.03± 0.00	0.03± 0.02	70.61± 0.78	98.52± 0.88	0.17± 0.00	0.21± 0.00
February	51.50± 0.20	65.10± 0.26	2.00± 0.05	1.65± 0.02	11.75± 0.50	12.86± 0.25	54.00± 0.00	46.28± 0.02	62.50± 0.15	47.08± 0.02	9.52± 1.37	18.24± 0.00	23.64± 0.02	32.85± 0.02	2.17± 0.04	2.14± 1.00	0.02± 0.00	0.08± 0.18	125.82± 0.15	65.34± 0.00	0.32± 0.00	0.36± 0.12
March	87.50± 0.51	86.15± 0.26	2.20± 0.00	2.14± 0.01	8.27± 3.01	9.65± 2.16	45.20± 0.02	32.00± 0.02	47.50± 4.50	62.05± 0.02	10.23± 0.24	25.00± 0.05	18.03± 0.00	25.06± 0.05	2.36± 0.85	2.35± 1.02	0.02± 0.00	0.02± **0.02**	185.73± 0.00	33.77± 0.15	0.38± 0.15	0.35± 0.05
April	66.25± 2.05	81.08± 0.05	2.37± 0.06	2.32± 0.03	12.86± 0.30	**6.35±** 0.10	49.50± 0.05	62.08± 0.02	40.00± 0.10	52.82± 0.05	15.36± 0.06	24.25± 0.12	15.82± 0.00	35.05± 0.02	2.72± 1.65	2.86± 0.81	0.04± 0.00	0.05± 0.00	**185.76±** 0.00	131.51± 0.00	0.27± 0.00	0.56± 0.03
May	85.90± 0.08	65.06± 0.05	2.53± 0.18	2.68± 0.02	14.70± 0.80	15.25± 0.30	50.71± 0.19	52.00± 0.05	66.26± 0.12	62.00± 0.12	17.94± 0.07	22.01± 0.00	15.06± 0.73	28.06± 0.31	2.70± 0.82	2.15± 0.02	0.05± 0.00	0.076± 0.02	211.35± 3.36	116.48± 1.27	0.52± 0.021	0.58± 0.06
June	57.50± 0.18	56.0± 0.05	5.68± 0.74	2.97± 0.12	5.32± 0.24	4.40± 0.02	54.90± 0.05	62.50± 0.12	41.25± 0.20	36.80± 0.03	20.23± 0.26	25.05± 0.35	19.39± 0.15	20.85± 0.08	2.15± 0.00	2.84± 1.00	0.05± 0.00	0.08± 0.05	221.91± 1.22	202.64± 0.19	0.62± 0.04	0.63± 0.18
July	70.35± 0.05	52.80± 2.15	3.44± 0.07	3.16± 0.52	4.51± 0.53	5.68± 0.21	76.25± 0.46	52.06± 0.08	50.00± 0.05	25.50± 0.60	19.95± 0.12	26.08± 0.02	19.09± 0.16	15.26± 0.06	1.80± 0.02	1.25± 0.00	0.58± 0.00	0.36± 0.00	278.11± 1.46	179.05± 0.10	0.76± 0.05	0.72± 0.00
August	80.40± 0.09	48.25± 2.16	3.75± 0.07	2.38± 0.03	5.10± 0.66	8.15± 0.14	98.00± 0.51	162.00± 0.34	71.25± 0.26	52.50± 0.02	19.83± 0.59	20.35± 0.16	22.65± 0.25	17.05± 0.10	1.32± 0.12	1.16± 0.02	0.67± 0.00	0.33± 0.03	366.61± 1.27	705.24± 1.17	0.64± 0.17	0.53± 0.05
September	73.75± 0.78	58.16± 7.16	3.80± 0.28	2.52± 0.62	5.16± 2.11	6.25± 0.16	122.50± 1.16	146.00± 0.21	68.17± 0.88	42.25± 0.12	15.00± 0.19	18.30± 0.12	25.17± 0.13	18.21± 0.24	1.64± 0.00	1.21± 0.00	0.06± 0.00	0.06± 0.05	473.55± 5.01	621.75± 0.15	0.56± 0.02	0.69± 0.20
October	75.83± 3.52	84.21± 0.26	4.10± 0.04	2.31± 0.10	6.38± 1.03	5.16± 2.02	56.20± 0.00	54.00± 0.05	36.50± 0.00	36.08± 0.03	11.22± 0.05	17.20± 0.21	27.19± 0.24	32.00± 0.02	1.25± 0.00	1.65± 0.02	0.04± 0.00	0.03± 0.03	87.63± 1.07	107.04± 0.15	0.52± 0.00	0.48± 0.02
November	17.25± 0.05	85.68± 2.16	2.86± 0.13	2.02± 0.10	7.22± 0.12	3.46± 0.03	86.00± 0.14	**78.00±** 0.02	47.58± 0.00	42.25± 0.04	9.16± 0.18	14.06± 0.01	30.16± 0.19	36.52± 0.05	0.15± 1.05	0.10± 0.00	0.04± 0.00	0.02± 0.02	271.68± 0.24	201.82± 0.15	0.37± 0.00	0.32± 0.02
December	25.22± 0.17	82.00± 0.05	3.71± 0.16	2.15± 0.06	13.62± 0.50	12.36± 0.24	65.00± 0.40	75.06± 0.16	65.32± 0.00	42.00± 0.04	8.46± 0.18	15.62± 0.16	35.17± 1.18	32.16± 0.02	2.54± 0.82	2.62± 0.12	0.27± 0.00	0.03± 0.00	95.99± 5.50	208.73± 0.68	0.20± 0.01	0.21± 0.01
Annual Average	60.14± 0.64	**70.03±** 1.40	3.33± 0.15	2.37± 0.14	8.95± 0.84	8.82± 0.47	68.06± 0.25	74.19± 0.11	54.69± 0.52	46.12± 0.10	13.72± 0.28	19.87± 0.10	23.96± 0.26	28.42± 0.09	1.93± 0.45	1.89± 0.34	0.16± 0.00	0.10± 0.04	214.56± 1.67	222.66± 0.41	0.44± 0.05	0.47± 0.06

Sd±Standard Deviation

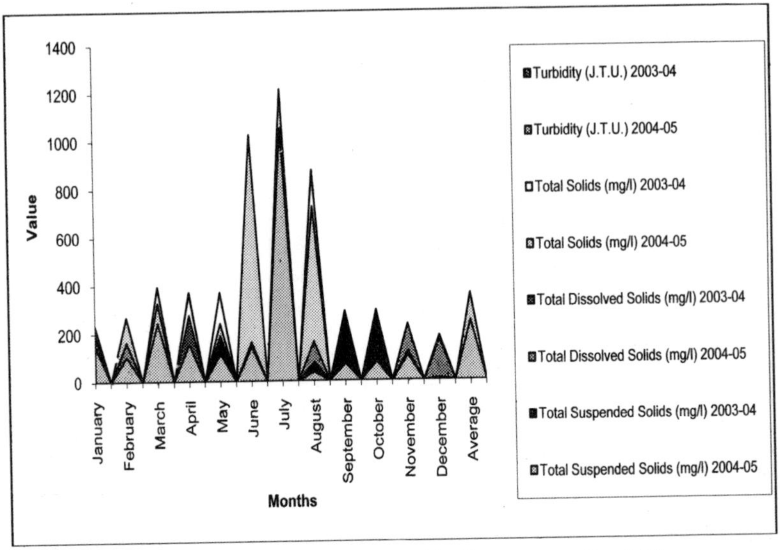

Fig. 4a.26 : Graph showing monthly fluctuation of turbidity, T.S.S., T.S. and T.D.S at Sampling station E

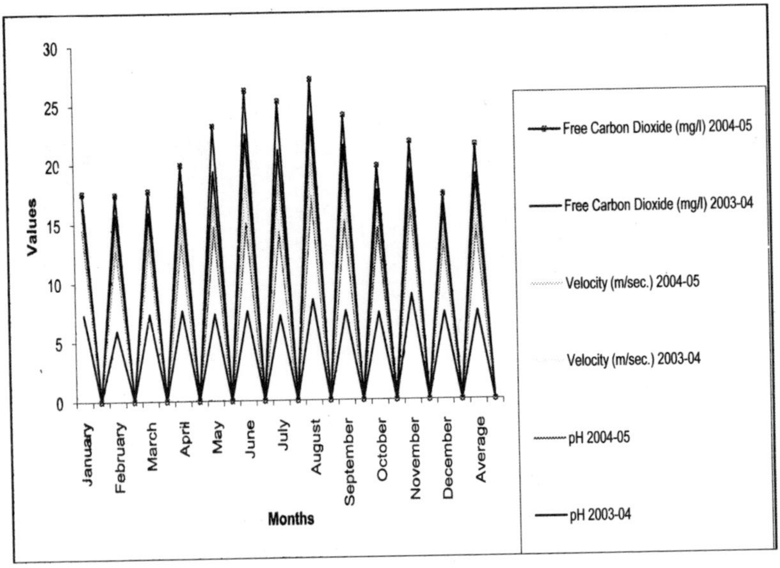

Fig. 4a.27 : Graph showing monthly fluctuation of Velocity, Free CO_2 and pH at Sampling station E

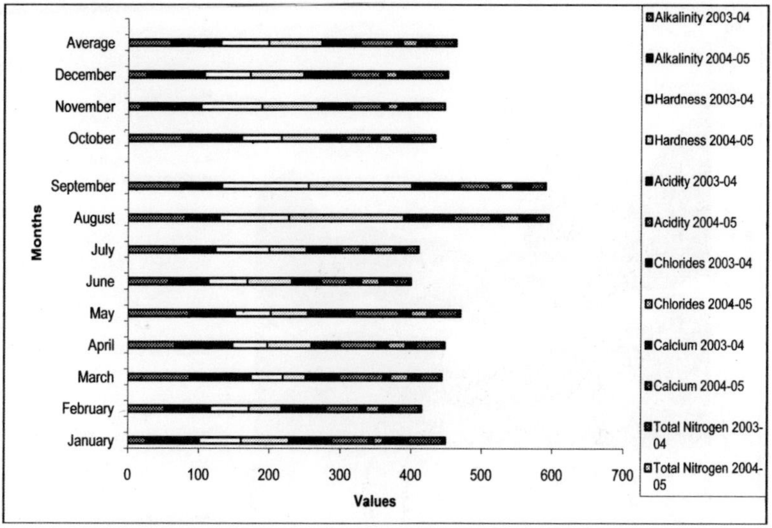

Fig. 4a.28: Graph showing monthly fluctuation of various chemical parameters at Sampling station E

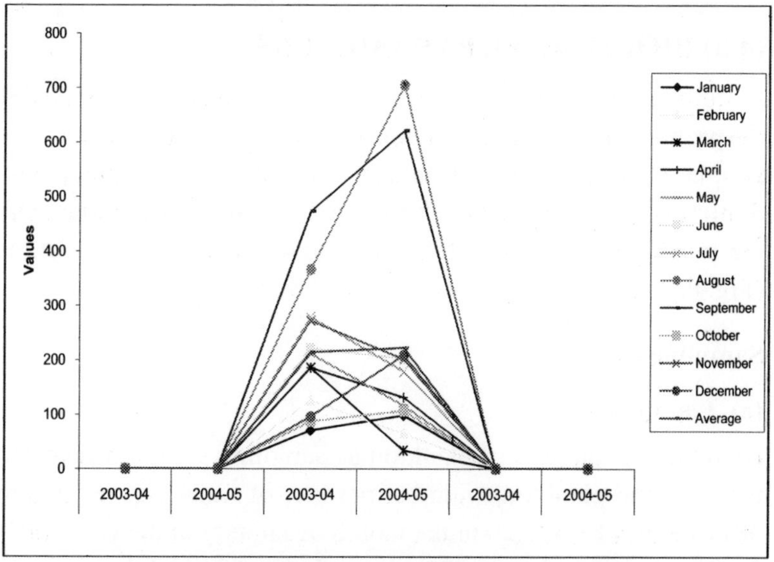

Fig. 4a.29: Graph showing monthly fluctuation of Cu, Mg and Fe at Sampling station E

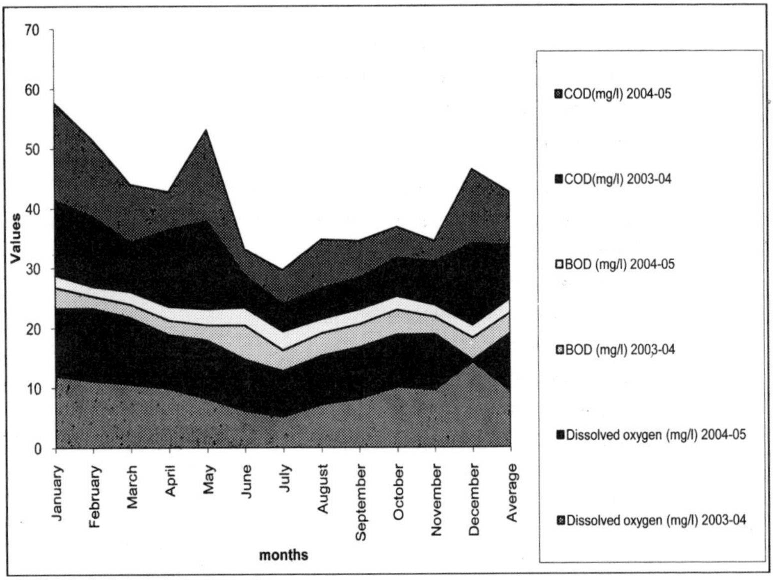

Fig. 4a.30: Graph showing monthly fluctuation of COD, BOD and DO at Sampling station E

(4.b) BIOLOGICAL PARAMETERS

A study on the biological parameters of the river Ganga includes the primary producers like phytoplankton, Primary Consumers like zooplankton, Tertiary Consumers like fishes and benthicfauna. Station wise details of the observations of biological parameters are mentioned in tables (Table 4b.1 to Table 4b.20) and figures (Fig.4b.1 to Fig. 4b.25)

Sampling station A

(a) Plankton :

Quantitative analysis of plankton at Sampling station A revealed that the total planktonic number per liter of water was maximum 3817.68 unit/l ± 477.20 in the month of January in the year 2004-2005 and minimum was 423.94 unit/l ± 209.82 in July month of the 2004-2005. (Table 4b.1-4b.3 and Figure 4b.1-4b.4)

1. Phytoplankton :

Phytoplankton was noted lowest of 362.28 unit/l ± 65.76 in the month of November of 2004-2005 and highest of 3297.70 unit/l ± 376.98 in January month of 2004-2005 with the average cells number was 1359.60 unit/l ± 280.06 in 2003-2004 and 1275.57 unit/l ± 185.88 in 2004-2005.

In phytoplankton Diatoms was dominated and class Blue green algae was found least in both the years of study. At this sampling station Diatoms was observed maximum in the month of January of 2003-2004 (2846.30 unit/l ± 720.00) and minimum in December month (129.65 unit/l ± 727.82) of 2003-2004 with average count of cells in 2003-2004 was 1151.77 unit/l ± 229.27 and 1087.14 unit/l ± 130.22 in 2004-2005. Green algae were found to be highest in the month of January 2003-2004 (364.86 unit/l ± 18.36) and lowest in 2004-2005 in the month of December (14.15unit/l ± 3.16). The average count of Green algae in 2003-2004 was 171.20 unit/l ± 33.48 and 142.87 unit/l ± 44.72 in 2004-2005. Blue green algae showed optimum peak during the month of January in 2004-2005 (162.36 unit/l ± 34.78) and least peak during the month of July of 2003-2004 (2.67unit/l ± 5.64). In 2003-2004 the average value of Blue green algae was 36.63 unit/l ± 17.31 and in 2004-2005 it was 45.56 unit/l ± 10.94. (Table 4b.2 and Fig. 4b.2 to Fig. 4b.3).

2. Zooplankton :

Zooplankton was observed minimum in July of 2004-2005 (19.00 unit/l ± 17.46) and maximum in 2004-2005 for the month of January (519.98 unit/l ± 100.22). and the average value of 219.60 unit/l ± 51.87 in 2003-2004 and in 2004-2005 the average was 214.91 unit/l ± 38.57.(Table 4b.1 and Table 4b.4) and (Fig. 4b.1 and Fig. 4b.4).

At this spot Keretala valga was maximum in the month of January in 2004-2005 (176.24 unit/l ± 24.36) and minimum was found 1.08-unit/l ± 5.64 in June month (2004-2005). Ceriodapnia was found to be highest in the month of January 2004-2005 (186.35 unit/l ± 25.40) and lowest (5.26 unit/l ± 6.16) in the month of July of the same year (2004-2005). Arcella was observed in their optimum peak during 2004-2005 March (112.16 unit/l ± 8.24) and in lowest peak during the month of August (1.00 unit/l ± 0.00) in the year 2003-2004. Crustacea was maximum (106.00 unit/l ± 15.26) in 2003-2004 February and minimum was found in the month of July 2003-2004 (8.12 unit/l ± 2.62). *Fillinia* sp. was observed in their optimum peak during 2003-2004 March (106.00 unit/l ± 32.05) and in lowest peak during the month of July (3.10unit/l ± 0.10) in the year 2003-2004. (Table 4b.1 and Figure 4b.1).

The average value revealed that Keretala valga was dominating in the zooplankton, 2003-2004 (61.11unit/l ± 13.53) followed by Ceriodapnia (56.08 unit/l ± 12.07). Crustacea (52.01 unit/l ± 13.01), *Fillinia* sp. (38.84 unit/l ± 9.61), *Arcella* (11.56 unit/l ± 3.22). In second year 2004-2005 Ceriodapnia (77.96 unit/l ± 11.43) found in strength followed by *Keretala valga* (55.65 unit/l ± 9.86), *Fillinia* sp. (43.76 unit/l ± 11.10) and *Arcella* (37.51 unit/l ± 6.19). (Table 4b.3 and Figure 4b.4).

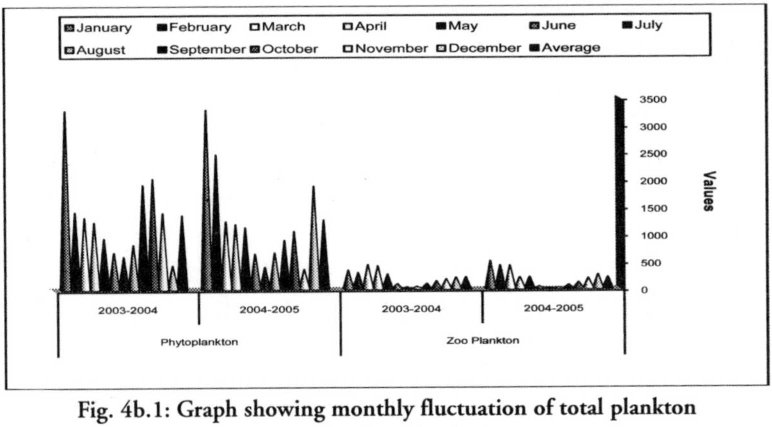

Fig. 4b.1: Graph showing monthly fluctuation of total plankton at Sampling station A

Table 4b.1: Quantitative Analysis of the Plankton of the River Ganga During 2003-2005 at sampling station A

Months	Number of Plankton per litre of water				Total Plankton	
	Phytoplankton		Zoo Plankton			
	2003-2004	2004-2005	2003-2004	2004-2005	2003-2004	2004-2005
January	3287.46 ± 808.36	3297.70±376.98	347.85±78.66	519.98±100.22	3635.31±887.02	3817.68±477.20
February	1425.88 ± 224.60	2470.34±232.63	304.14±47.95	439.17±67.18	1730.02±272.55	2909.51±299.81
March	1322.89 ± 158.08	1247.33±182.42	450.90±117.63	437.56±41.03	1773.79±275.71	1684.89±223.45
April	1237.31 ± 266.89	1192.22±201.99	425.11±80.36	214.76±39.36	1662.42±347.25	1406.98±241.35
May	939.85 ± 111.60	1138.35±172.29	274.47±58.96	221.42±28.45	1214.32±170.56	1359.77±200.74
June	681.37 ± 125.26	654.29±197.30	97.15±29.01	47.31±21.19	778.52±154.27	701.60±218.49
July	601.42 ± 51.91	404.94±192.36	40.03±14.80	19.00±17.46	641.45±66.71	423.94±209.82
August	821.98 ± 117.70	674.42±122.25	49.68±19.62	24.68±26.32	1071.66±137.32	699.10±148.57
September	1920.84 ± 208.44	903.25±157.60	103.23±36.04	74.07±25.05	2024.07±244.48	977.32±182.65
October	2033.40 ± 193.11	1067.24±94.51	146.82±34.23	121.51±31.79	2180.22±227.34	1188.75±126.30
November	1405.78 ± 180.08	362.28±65.76	181.83±40.19	195.87±26.32	1587.61±220.27	558.15±92.08
December	437.00 ± 914.67	1894.48±234.46	214.03±65.01	263.53±38.52	651.03±979.68	2158.01±272.98
Average	**1359.60 ± 280.06**	**1275.57±185.88**	**219.60±51.87**	**214.91±38.57**	**1579.20±331.93**	**1490.48±224.45**

Sd±Standard Deviation

Table 4b.2: Number and Percentage of different Group among the Phytoplankton of the River Ganga During 2003-2005 at sampling station A

Months	Total Phytoplankton/ liter of water		Total Diatoms / Litre of water		Total green algae/ Litre of water		Total Blue green algae/ litre of water	
	2003-04	2004-05	2003-04	2004-05	2003-04	2004-05	2003-04	2004-05
January	3287.46 ±808.36	3297.70± 376.98	2846.30 ±720.00	2783.24 ±125.20	364.86 ±18.36	352.10 ±217.00	76.30 ±70.00	162.36± 34.78
February	1425.88 ±224.60	2470.34± 232.63	1100.62 ±178.36	2106.52 ±156.38	242.62 ±28.04	242.52 ±26.15	82.64 ±18.20	121.30± 50.10
March	1322.89 ±158.08	1247.33± 182.42	1098.89 ±136.82	1048.37 ±136.85	184.00 ±12.62	162.18 ±43.16	40.00 ±8.64	36.78 ±2.41
April	1237.31 ±266.89	1192.22± 201.99	975.12 ±234.00	1015.22 ±172.62	186.83 ±18.29	135.00 ±25.00	75.36 ±14.60	42.00 ±4.37
May	939.85 ±111.60	1138.35± 172.29	836.72 ±48.95	1010.00 ±154.31	72.35 ±42.79	106.15 ±12.72	30.78 ±19.86	22.20 ±5.26
June	681.37 ±125.26	654.29± 197.30	610.00 ±78.29	562.48 ±178.06	66.37 ±40.25	85.65 ±17.18	5.00 ±6.72	6.16 ±2.06
July	601.42 ±51.91	404.94± 192.36	546.37 ±35.62	382.52 ±187.52	52.38 ±10.65	14.15 ±3.16	2.67 ±5.64	8.27 ±1.68
August	821.98 ±117.70	674.42± 122.25	970.45 ±87.96	654.17 ±115.24	45.82 ±19.86	15.10 ±5.15	5.71 ±9.88	5.15 ±1.86
September	1920.84 ±208.44	903.25± 157.60	1752.00 ±168.75	782.15 ±132.36	156.20 ±22.80	95.68 ±17.10	12.64 ±16.89	25.42 ±8.24
October	2033.40 ±193.11	1067.24± 94.51	1820.00 ±172.65	897.24 ±16.56	182.90 ±6.78	125.00 ±62.71	30.50 ±13.68	45.00 ±15.24
November	1405.78 ±180.08	362.28± 65.76	1135.10 ±162.00	182.20 ±27.40	268.04 ±9.48	164.30 ±35.20	2.64 ±8.60	15.78 ±3.16
December	437.00 ±914.67	1894.48± 234.46	129.65 ±727.82	1621.56 ±160.25	232.00 ±171.86	216.65 ±72.15	75.35 ±14.99	56.27 ±2.06
Average	**1359.60 ±280.06**	**1275.57± 185.88**	**1151.77 ±229.27**	**1087.14 ±130.22**	**171.20 ±33.48**	**142.87 ±44.72**	**36.63 ±17.31**	**45.56 ±10.94**

Sd±Standard Deviation

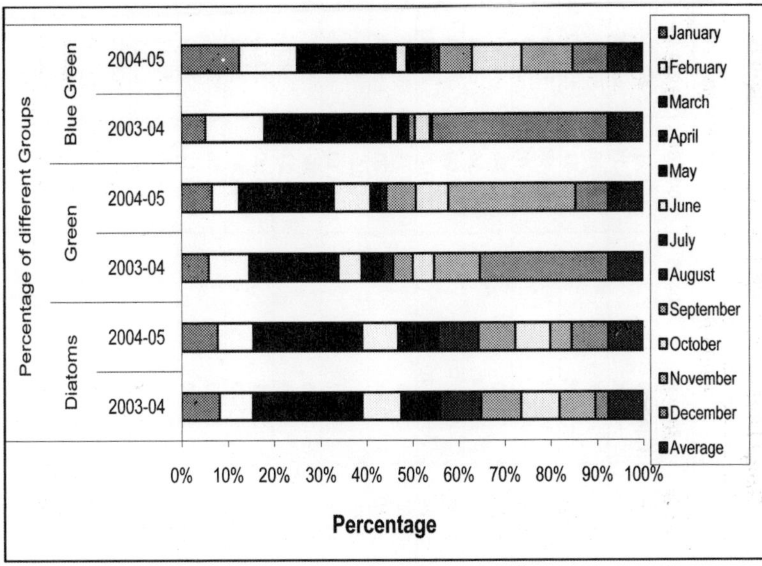

Fig. 4b.2: Graph showing percentage of different groups of phytoplankton at Sampling station A

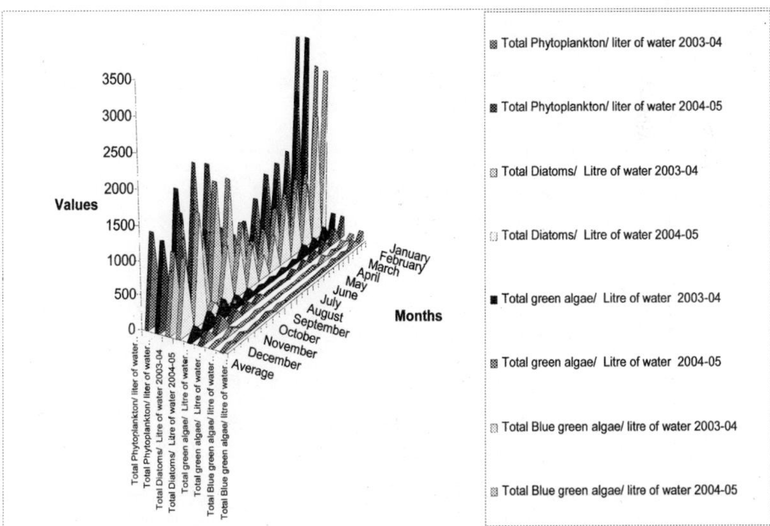

Fig. 4b.3: Graph showing values of different groups phytoplankton at Sampling station A

Table 4b.3: Number of Different Genera among Zooplankton in the River Ganga During 2003-2005 at sampling station A

Months	*Keretala valga*		*Ceriodapnia*		*Arcella*		*Crustaclea*		*Fillinia sp.*	
	2003-2004	2004-2005	2003-04	2004-05	2003-04	2004-05	2003-04	2004-05	2003-04	2004-05
January	105.00 ±24.36	176.24 ±24.36	78.50 ±13.00	186.35 ±25.40	15.10 ±6.14	52.15 ±18.46	85.00 ±20.00	-	64.25 ±15.16	105.24 ±32.00
February	85.54 ±13.52	106.56 ±17.24	64.24 ±6.24	156.21 ±18.26	26.20 ±6.12	78.24 ±15.16	106.00 ±15.26	-	25.16 ±6.81	98.16 ±16.52
March	124.52 ±37.25	112.00 ±12.17	86.00 ±17.21	135.15 ±12.35	32.00 ±6.00	112.16 ±8.24	102.38 ±25.12	-	106.00 ±32.05	78.25 ±8.27
April	108.65 ±17.04	46.24 ±8.27	106.00 ±15.12	120.00 ±18.67	15.61 ±3.20	32.12 ±6.10	98.00 ±36.14	-	96.85 ±8.86	16.40 ±6.32
May	82.15 ±12.31	55.16 ±7.36	85.00 ±20.14	95.24 ±8.68	10.00 ±2.10	35.78 ±3.24	62.32 ±14.21	-	35.00 ±10.20	35.24 ±9.17
June	52.24 ±15.20	1.08 ±5.64	15.62 ±3.68	15.15 ±7.21	2.15 ±1.24	15.62 ±2.10	12.00 ±2.64	-	15.14 ±6.25	15.46 ±6.24
July	18.00 ±8.56	5.24 ±3.35	6.81 ±2.52	5.26 ±6.16	4.00 ±1.00	3.26 ±1.80	8.12 ±2.62	-	3.10 ±0.10	5.24 ±6.15
August	25.64 ±7.25	6.17 ±8.10	8.14 ±6.00	8.00 ±8.10	1.00 ±0.00	5.35 ±1.86	12.36 ±5.10	-	2.54 ±1.27	5.16 ±8.26
September	35.00 ±8.16	2.21 ±7.32	32.00 ±15.16	30.62 ±7.89	5.62 ±2.54	21.24 ±3.52	15.61 ±3.76	-	15.00 ±6.42	20.00 ±6.32
October	20.20 ±2.46	15.42 ±8.61	62.00 ±12.43	45.26 ±8.00	15.36 ±7.12	25.62 ±5.86	24.20 ±4.10	-	25.06 ±8.12	35.21 ±9.32
November	38.52 ±3.56	46.35 ±7.40	72.15 ±12.36	52.05 ±8.15	3.16 ±1.21	52.35 ±4.62	36.00 ±15.06	-	32.00 ±8.00	45.12 ±6.15
December	40.80 ±12.65	95.16 ±8.52	56.52 ±21.00	86.27 ±8.27	8.56 ±2.00	16.24 ±3.27	62.15 ±17.26	-	46.00 ±12.10	65.86 ±18.46
Average	**61.11 ±13.53**	**55.65 ±9.86**	**56.08 ±12.07**	**77.96 ±11.43**	**11.56 ±3.22**	**37.51 ±6.19**	**52.01 ±13.44**	-	**38.84 ±9.61**	**43.78 ±11.10**

Sd±Standard Deviation

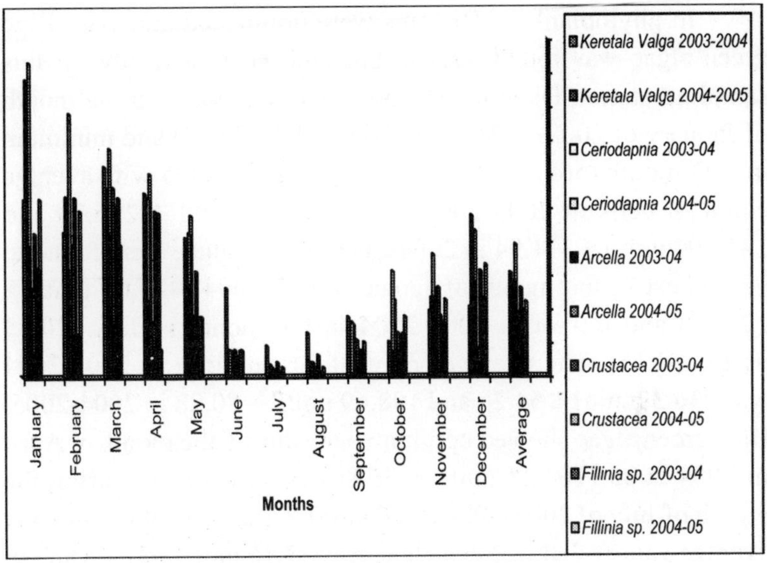

Fig. 4b.4: Graph showing values of different groups Zooplankton at Sampling station

Sampling station B

(a) Plankton:

Quantitative analysis of plankton at Sampling station B revealed that the total Planktonic number per litre of water was maximum 4849.29 unit/l ± 575.48 in the month of February in the year 2004-2005 and minimum was 700.63 unit/l ± 144.20 in July month of the 2004-2005. (Table 4b.4-Table 4b.6 and Figure 4b.5- Figure 4b.8)

1. Phytoplankton :

Phytoplankton was noted lowest of 659.03 unit/l ± 64.00 in the month of July of 2004-2005 and highest of 4486.80 unit/l ± 258.08 in January month of 2004-2005 with the average cells number was 2207.69unit/l ± 329.62 in 2003-2004 and 1693.72 unit/l ± 223.60 in 2004-2005. (Table 4b.4 and Figure 4b.5)

In phytoplankton Diatoms were dominated and class Blue green algae was found least in both the years of study. At this sampling station Diatoms was observed maximum in the month of January of 2003-2004 (4701.56 unit/l ± 157.03) and minimum in July month (568.32 unit/l ± 82.16) of 2004-2005 with average count of cells in 2003-2004 was 1924.22 unit/l ± 256.70 and 1445.94 unit/l ± 111.94 in 2004-2005. Green algae were found to be highest in the month of January 2003-2004 (480.02 unit/l ± 127.30) and lowest in 2003-2004 in the month of July (20.35 unit/l ± 3.15). The average count of Green algae in 2003-2004 was 236.32unit/l ± 54.26 and 198.69 unit/l ± 80.28 in 2004-2005. Blue green algae showed optimum peak during the month of April in 2004-2005 (164.25 unit/l ± 56.86) and least peak during the month of July of 2003-2004 (5.26 unit/l ± 6.03). In 2003-2004 the average value of Blue green algae was 47.16 unit/l ± 18.66 and in 2004-2005 it was 49.08 unit/l ± 31.38. (Table 4b.5 and Figure 4b.6 to Figure 4b.7).

2. Zooplankton :

Zooplankton were observed minimum in July of 2004-2005 (19.00 unit/l ± 17.46) and maximum in 2004-2005 for the month of January (519.98 unit/l ± 100.22). and the average value of 219.60 unit/l ± 51.87 in 2003-2004 and in 2004-2005 the average was 214.91 unit/l ± 38.57. (Table 4b.1 and Table 4b.4) and (Fig 4b.4 and Fig 4b.5).

At this spot *Keretala valga* was maximum in the month of January in 2003-2004 (182.63 unit/l ± 37.08) and minimum was found 6.57 unit/l ± 2.03 in July month (2003-2004). *Ceriodapnia* was found to be highest in the month of March 2003-2004 (149.00 unit/l ± 89.02) and lowest (12.00 unit/l ± 8.20) in the month of July of the same year (2003-2004). *Arcella* was observed in their optimum peak during 2003-2004 March (190.00 unit/l ± 32.61) and in lowest peak during the month of July (2.93 unit/l ± 0.24) in the year 2003-2004.*Crustacea* was maximum (135.27unit/l ± 42.31) in 2003-2004 March and minimum was found in the month of July 2003-2004 (3.17 unit/l ± 0.06). *Fillinia* sp. was observed in their optimum peak during 2003-2004 March (152.61 unit/l ± 35.49) and in lowest peak during the month of July (5.24unit/l ± 1.40) in the year 2004-2005. (Table 4b.6 and Figure 4b.8).

The average value revealed that *Keretala valga* was dominating in the zooplankton, 2003-2004 (78.40 unit/l ± 19.86) followed by *Arcella* (76.97 unit/l ± 20.22). *Ceriodapnia* (70.34 unit/l ± 27.46), *Fillinia* sp. (51.63 unit/l ± 15.23) *Crustacea* (50.31 unit/l ± 11.93). In second year 2004-2005 *Keretala valga* (61.41 unit/l ± 15.27) found in strength followed by *Ceriodapnia* and *Arcella* (54.28unit/l ± 12.45) and Fillinia sp. (40.13 unit/l ± 11.37). (Table 4b.6 and Figure 4b.8).

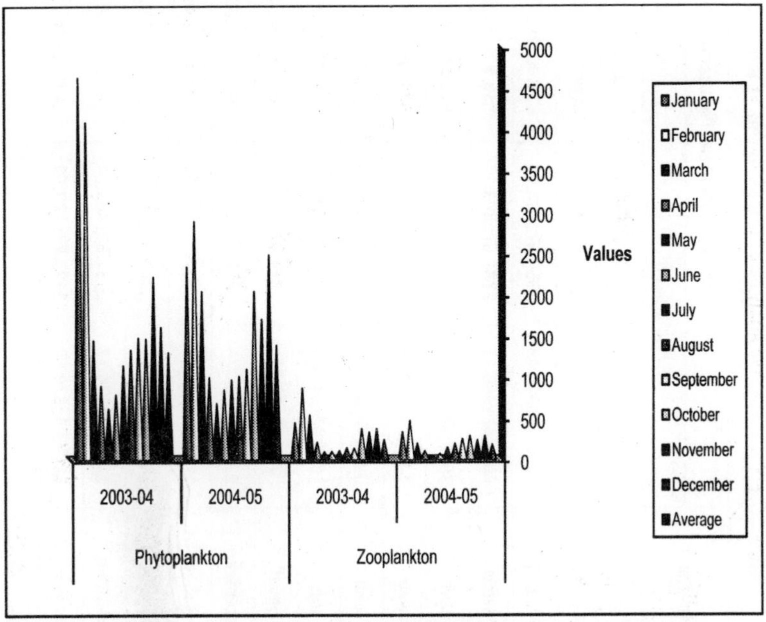

Fig. 4b.6: Graph showing monthly fluctuation of total plankton at Sampling station B

Table 4b.4: Quantitative analysis of the plankton of the river Ganga during 2003-2005 at Sampling Station B

Months	Number of Plankton Per Liter of Water				Total Plankton	
	Phytoplankton		Zooplankton			
	2003-04	2004-05	2003-04	2004-05	2003-04	2004-05
January	4628.02±1381.92	2329.08±398.73	436.52±35.60	323.56±1260.20	5064.54±141.75	2652.64±1658.93
February	5166.47±300.43	2954.57±422.00	320.00±160.00	362.49±317.40	5486.47±460.43	3317.06±739.4
March	4085.41±340.21	2884.87±309.11	856.62±132.00	460.07±158.48	4922.03±472.21	3344.64±467.59
April	3508.19±291.61	2544.57±233.40	621.34±158.62	320.12±168.80	4129.53±450.23	2864.69±402.2
May	1440.25±702.29	2029.67±202.46	526.10±127.80	184.75±433.10	1966.35±830.09	2214.42±635.53
June	887.00±100.11	989.76±213.86	196.25±72.36	92.93±184.69	1083.25±172.47	1082.69±398.55
July	604.92±34.40	670.94±112.76	84.60±35.00	41.6±80.20	689.52±69.40	712.54±192.96
August	781.75±26.21	839.34±57.76	87.21±18.24	59.6±49.49	868.96±44.45	898.94±107.25
September	1136.10±221.74	959.07±97.14	95.27±31.00	136.39±110.65	1231.37±252.74	1095.46±207.79
October	1324.18±136.26	1002.73±113.74	132.16±56.06	183.7±68.83	1456.29±192.32	1186.43±182.57
November	1471.87±146.14	1088.38±81.55	125.00±7.86	240.99±90.67	1596.87±154	1329.37±172.22
December	1458.17±274.17	2031.62±440.74	362.00±54.67	275.67±163.14	2569.69±384.27	2307.29±203.88
Average	2207.69±329.62	1693.72±223.60	320.25±74.10	223.48±257.13	2588.73±302.03	1920.25±447.82

Sd±Standard Deviation

Table 4b.5: Number and Percentage of different Group among the Phytoplankton of the River Ganga During 2003-2005 at sampling station B

Months	Total Phytoplankton/ Liter of water		Total Diatoms / Liter of water		Total green algae/ Liter of water		Total Blue green algae/ Liter of water		Percentage of different Groups					
									Diatoms		Green		Blue Green	
	2003-04	2004-05	2003-04	2004-05	2003-04	2004-05	2003-04	2004-05	2003-04	2004-05	2003-04	2004-05	2003-04	2004-05
January	4628.02 ±1381.92	2329.08 ±398.73	4073.00 ±1195.51	1873.56 ±162.00	480.02 ±127.30	369.00 ±172.46	75.00 ±59.11	86.52 ±64.27	85.66	80.44	10.94	15.84	2.11	3.71
February	5166.47 ±300.43	2954.57± 422.00	4701.56 ±157.03	2621.24 ±172.00	397.67 ±125.30	256.76 ±182.00	67.24 ±18.03	76.57 ±68.00	97.71	88.72	6.71	8.69	2.09	2.59
March	4085.41 ±340.21	2884.87± 309.11	3705.85 ±239.86	2562.62 ±178.64	311.00 ±97.60	240.00 ±97.80	68.56 ±2.75	82.25 ±32.67	92.53	88.83	8.53	8.32	2.00	2.85
April	3508.19 ±291.61	2544.57± 233.40	3170.52 ±247.17	2165.30 ±120.46	274.67 ±27.07	215.02 ±56.08	63.00 ±17.37	164.25± 56.86	85.00	85.09	14.25	8.45	4.35	6.45
May	1440.25 ±702.29	2029.67± 202.46	1152.00± 574.24	1781.39 ±168.56	232.00 ±97.67	200.00 ±17.08	56.25 ±30.38	48.28 ±16.82	81.69	87.77	25.00	9.85	2.79	2.38
June	887.00 ±100.11	989.76± 213.86	805.00± 67.50	868.56 ±125.40	75.60 ±17.30	115.00 ±16.00	6.40 ±15.31	6.20 ±72.46	89.76	87.75	9.75	11.62	1.86	0.63
July	604.92 ±34.40	670.94± 112.76	579.31± 25.22	568.32 ±82.16	20.35 ±3.15	96.00 ±12.74	5.26 ±6.03	6.62 ±17.86	92.20	84.71	9.07	14.31	1.70	0.99
August	781.75 ±26.21	839.34± 57.76	739.56± 17.60	732.00 ±35.62	35.09 ±4.59	98.72 ±13.52	7.10 ±4.02	8.62 ±8.62	90.81	87.21	6.75	11.76	1.26	1.03
September	1136.10 ±221.74	959.07± 97.14	913.50± 167.74	814.00 ±65.00	170.25 ±23.00	125.07 ±26.00	52.35 ±31.00	20.00 ±6.14	57.50	84.87	12.72	13.04	6.54	2.09
October	1324.18 ±136.26	1002.73± 113.74	1075.54± 105.09	826.56 ±38.64	189.00 ±15.73	164.00 ±72.00	59.64 ±15.44	12.17 ±3.10	78.62	82.43	11.54	16.36	8.17	1.21
November	1471.87 ±146.14	1088.38± 81.55	1159.76± 112.92	856.24 ±62.25	274.55 ±26.70	226.72 ±15.68	37.56 ±6.52	5.42 ±3.62	65.93	78.67	25.00	20.83	3.24	0.50
December	1458.17 ±274.17	2031.62± 440.74	1015.00± 170.54	1681.52 ±132.60	375.67 ±85.63	278.00 ±282.00	67.50 ±18.00	72.10 ±26.14	58.17	82.77	12.75	13.68	2.50	3.55
Average	**2207.69 ±329.62**	**1693.72± 223.60**	**1924.22± 256.70**	1445.94 ±111.94	236.32 ±54.26	198.69 ±80.28	47.16 ±18.66	49.08 ±31.38	**81.30**	**84.94**	**12.75**	**12.73**	**3.22**	**2.33**

Sd±Standard Deviation

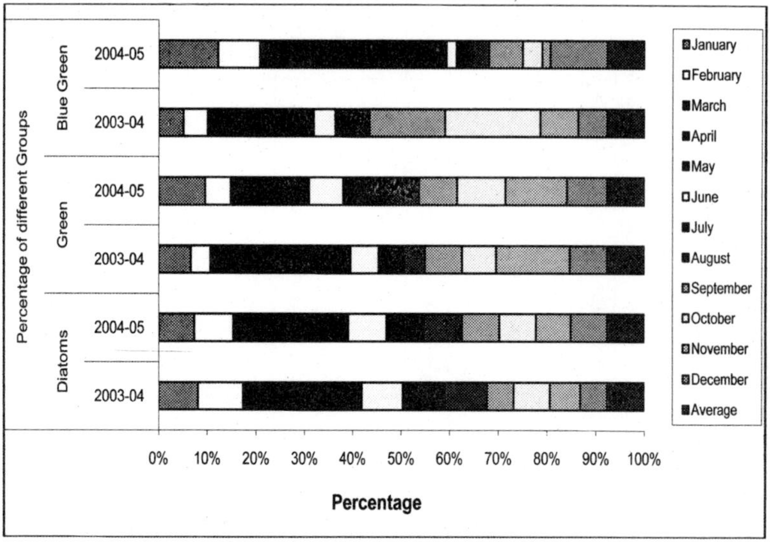

Fig. 4b.7: Graph showing percentage of different groups of phytoplankton at Sampling station B

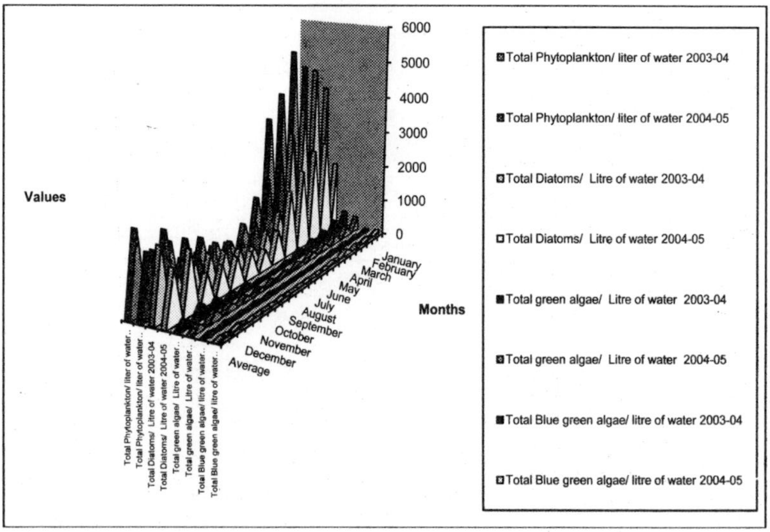

Fig. 4b.8: Graph showing values of different groups phytoplankton at Sampling station B

Table 4b. 6: Number of Different Genera among Zooplankton in the River Ganga During 2003-2005 at sampling station B

Months	***Keretala valga***		***Ceriodaphnia***		***Arcella***		***Crustacea***		***Fillinia sp.***	
	2003-04	**2004-05**	**2003-04**	**2004-05**	**2003-04**	**2004-05**	**2003-04**	**2004-05**	**2003-04**	**2004-05**
January	182.63 ±37.08	86.21 ±7.26	106.00 ±32.71	112.40 ±26.56	137.00 ±38.51	112.40 ±26.56	86.00 ±12.37	-	115.67 ±42.15	65.66 ±17.60
February	157.00 ±42.00	108.25 ±68.20	115.64 ±61.00	100.16 ±17.86	189.68 ±42.80	112.00 ±24.16	64.30 ±16.20	-	121.03 ±24.00	70.61 ±16.25
March	172.65 ±41.38	105.40 ±18.16	149.00 ±89.02	112.51 ±18.42	190.00 ±32.61	68.71 ±16.15	135.27 ±42.31	-	152.61 ±35.49	57.10 ±25.00
April	110.36 ±10.53	100.67 ±28.62	107.20 ±12.86	106.00 ±24.10	172.53 ±32.06	85.66 ±23.26	85.62 ±5.82	-	83.00 ±6.52	40.68 ±8.62
May	74.21 ±50.24	86.52 ±12.71	78.35 ±42.79	56.68 ±17.82	52.62 ±28.40	55.00 ±16.21	53.71 ±28.80	-	52.85 ±40.48	32.10 ±2.46
June	14.00 ±6.28	56.14 ±11.24	21.71 ±6.28	24.60 ±8.16	14.68 ±2.98	12.16 ±6.24	14.39 ±5.26	-	24.30 ±10.32	10.67 ±8.60
July	6.57 ±2.03	25.32 ±6.10	12.00 ±8.20	15.61 ±3.16	2.93 ±0.24	10.25 ±3.16	3.71 ±0.06	-	9.00 ±1.02	5.24 ±1.40
August	10.16 ±3.53	15.12 ±2.14	15.52 ±3.42	16.00 ±4.08	10.26 ±0.09	17.86 ±2.14	5.28 ±1.35	-	13.53 ±4.12	17.86 ±8.25
September	32.62 ±7.28	16.25 ±6.16	30.25 ±7.18	25.62 ±17.86	7.16 ±1.20	32.52 ±8.67	17.87 ±6.24	-	28.02 ±3.35	25.60 ±7.30
October	34.00 ±8.05	25.30 ±7.86	42.80 ±5.24	32.10 ±16.10	32.80 ±13.42	52.16 ±6.16	28.35 ±4.20	-	28.65 ±2.10	40.75 ±5.66
November	52.62 ±12.39	36.10 ±6.10	68.40 ±8.13	45.80 ±15.42	49.63 ±17.68	57.00 ±8.10	36.42 ±4.89	-	50.00 ±0.81	50.63 ±7.08
December	94.00 ±17.52	75.64 ±8.67	97.24 ±52.64	86.30 ±8.26	64.32 ±32.65	35.62 ±8.62	72.82 ±15.63	-	60.89 ±12.39	64.66 ±17.60
Average	**78.40 ±19.86**	**61.41 ±15.27**	**70.34 ±27.46**	**54.28 ±12.45**	**76.97 ±20.22**	**54.28 ±12.45**	**50.31 ±11.93**	-	**61.63 ±15.23**	**40.13 ±11.37**

Sd±Standard Deviation

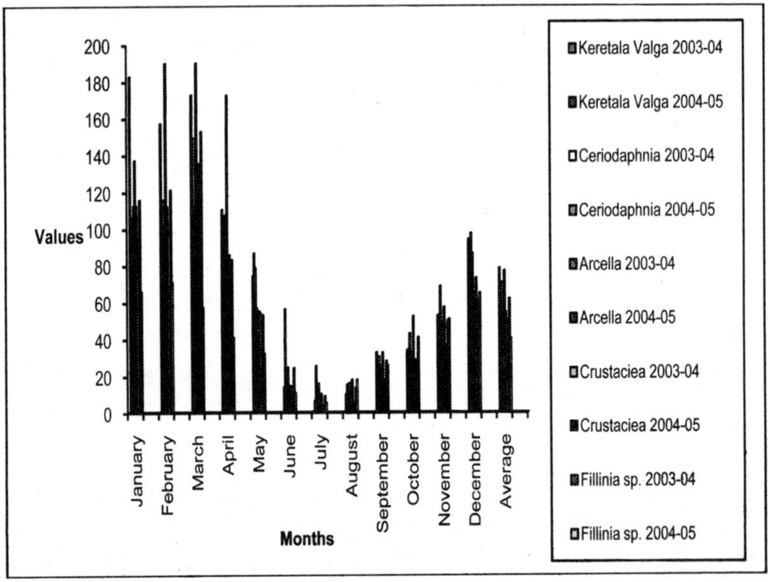

Fig. 4b.9: Graph showing values of different groups Zooplankton at Sampling station B

Sampling station C

(a) Plankton:

Quantitative analysis of plankton at Sampling station C revealed that the total Planktonic number per litre of water was maximum 9368.71 unit/l ± 457.29 in the month of January in the year 2003-2004 and minimum was 534.51 unit/l ± 122.98 in July month of the 2004-2005. (Table 4b.7- 4b.9 and Figure 4b.9-4b.12).

1. Phytoplankton :

Phytoplankton was noted lowest of 518.90 unit/l ± 110.56 in the month of July of 2004-2005 and highest of 8933.33 unit/l ± 300.62 in January month of 2003-2004 with the average cells number was 3545.49 unit/l ± 342.80 in 2003-2004 and 1639.03 unit/l ± 161.12 in 2004-2005. (Table 4b.7, Figure 4b.9)

In phytoplankton Diatoms was dominated and class Blue green algae was found least in both the years of study. At this sampling station Diatoms was observed maximum in the month of February of 2003-2004 (4152.23 unit/l ± 274.62) and minimum in July month (639.99 unit/l ± 72.10) of 2004-2005 with average count of cells in 2003-2004 was 1899.33 unit/l ± 228.06 and 1364.79 unit/l ± 121.17 in 2004-2005. Green algae were found to be highest in the month of January 2004-2005 (368.00 unit/l ± 15.00) and lowest in 2003-2004 in the month of September (12.16unit/l ± 18.15). The average count of Green algae in 2003-2004 was 158.30 unit/l ± 20.28 and 171.89 unit/l ± 11.25 in 2004-2005. Blue green algae showed optimum peak during the month of January in 2004-2005 (92.52 unit/l ± 12.00) and least peak during the month of July of 2003-2004 (4.32 unit/l ± 2.48). In 2003-2004 the average value of Blue green algae was 39.21 unit/l ± 8.80 and in 2004-2005 it was 35.51 unit/l ± 12.19 (Table 4b.8 and Figure 4b.10 to 4b.11).

2. Zooplankton :

Zooplankton was observed minimum in July of 2004-2005 (15.61 unit/l ± 12.42) and maximum in 2004-2005 for the month of January (545.29 unit/l ± 101.28) and the average value of 257.69 unit/l ± 87.93 in 2003-2004 and in 2004-2005 the average was 236.90 unit/l ± 48.02. (Table 4b.7 and Table 4b.9) and (Figure 4b.9 and Figure 4b.12).

At this spot *Keretala valga* was maximum in the month of January in 2003-2004 (149.65 unit/l ± 56.30) and minimum was found 10.20 unit/l ± 3.61 in July month. (2004-2005). *Ceriodapnia* was found to be highest in the month of March 2004-2005 (196.10 unit/l ± 25.00) and lowest (11.00 unit/l ± 4.62) in the month of July of the year 2003-2004. *Arcella* was observed in their optimum peak during 2003-2004 February (137.25 unit/l ± 58.82) and in lowest peak during the month of July (11.26 unit/l ± 3.12) in the year 2004-2005. *Fillinia* sp. was observed in their optimum peak during 2004-2005 January (172.00 unit/l ± 19.20) and in lowest peak during the month of July (6.14unit/l ± 0.68) in the year 2004-2005. (Table 4b.9 and Figure 4b.12).

The average value revealed that *Keretala valga* was dominating in the zooplankton, 2003-2004 (68.91unit/l ± 21.35) followed by *Arcella* (63.82 unit/l ± 16.55), *Fillinia* sp. (47.56 unit/l ± 29.20), *Ceriodapnia* (43.20 unit/l ± 26.03). In second year 2004-2005 *Ceriodapnia* (76.60 unit/l ± 16.32) found in strength followed by *Fillinia* sp. (52.05 unit/l ± 11.40), *Arcella* (50.12 unit/l ± 10.53) and *Keretala valga* (36.01 unit/l ± 8.61). (Table 4b.9 and Figure 4b.12).

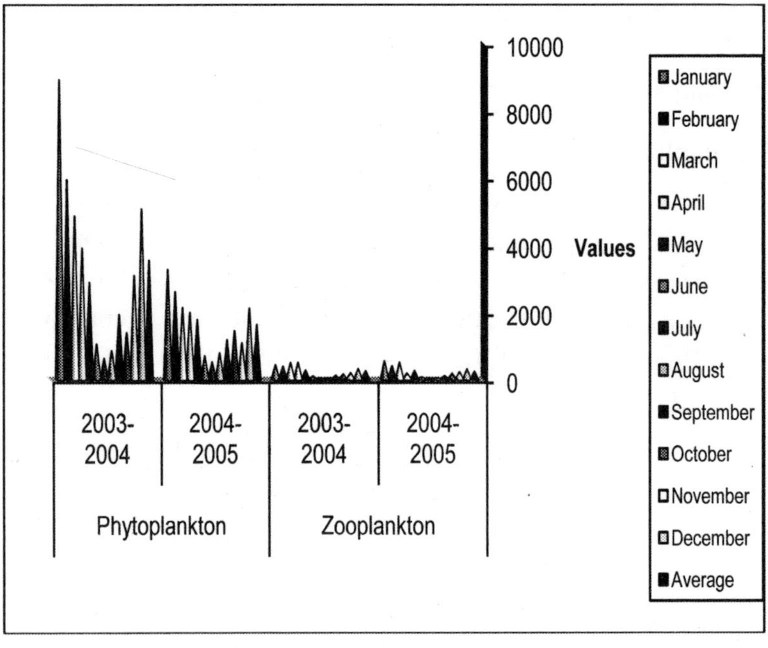

Fig. 4b.11: Graph showing monthly fluctuation of total plankton at Sampling station C

Table 4b.7: Quantitative Analysis of the Plankton of the River Ganga During 2003-2005 at Sampling Station C

Months	**Number of Plankton per liter of water**				**Total Plankton**	
	Phytoplankton		**Zooplankton**			
	2003-2004	**2004-2005**	**2003-2004**	**2004-2005**	**2003-2004**	**2004-2005**
January	3443.86±1260.20	2112.52±223.52	434.78±156.67	545.29±101.28	3878.64±1416.87	2657.81±324.8
February	4486.80±317.40	2596.35±279.51	390.89±141.65	405.67±75.17	4877.69±459.05	3002.02±354.68
March	3319.24±158.48	2544.81±208.72	503.56±244.84	509.67±98.28	3822.8±403.32	3054.48±307
April	2472.92±168.80	2325.52±296.28	507.89±74.74	182.83±35.93	2980.81±243.54	2508.35±332.21
May	2845.66±433.10	1853.00±191.67	279.55±132.74	265.16±55.98	3125.21±565.84	2118.16±247.65
June	1121.00±184.69	1093.19±48.38	100.36±43.77	69.99±23.46	1221.36±228.46	1162.18±71.84
July	659.03±80.20	710.35±137.07	29.69±15.11	15.61±12.42	688.72±95.31	725.96±149.49
August	854.22±49.49	761.04±46.18	37.64±21.04	31.66±17.06	891.86±70.53	792.7±63.24
September	1025.52±110.65	978.05±80.09	120.08±45.21	107.95±36.97	1145.6±155.86	1086±117.06
October	1237.19±68.83	1052.82±35.12	162.38±39.39	179.75±38.95	1399±108.22	1232.57±74.07
November	1225.09±90.67	1240.33±46.56	197.83±56.74	223.54±27.87	1422.92±147.42	1463.87±74.43
December	2471.62±163.14	1598.21±142.27	317.60±83.31	305.72±52.84	2843.22±246.45	1903.93±195.11
Average	**2096.85±257.14**	**1572.18±144.61**	**257.69±87.93**	**236.90±48.02**	**2358.2±345.07**	**1809.08±192.63**

Sd±Standard Deviation

Table 4b.8: Number and Percentage of different Group among the Phytoplankton of the River Ganga During 2003-05 at Sampling Station C

Months	Total Phytoplankton/ liter of water		Total Diatoms /Liter of Water		Total Green Algae/ Litre of Water		Total Blue Green algae/ Litre or Water		Percentage of different (Groups (%)		
	2003-04	**2004-05**	**2003-04**	**2004-05**	**2003-04**	**2004-05**	2003-04	2004-05	**Diatoms**	**Green**	**Blue Green**
January	3443.86 ±1260.20	2112.52± 223.52	3032.54± 1230.01	1210.00± 128.00	326.08± 25.65	368.00± 15.00	85.24± 4.54	92.52± 12.00	88.06 78.20	9.47 17.42	2.48 4.38
February	4486.80± 317.40	2596.35± 279.51	4152.23± 274.62	1652.00± 196.52	278.25± 26.08	242.15± 18.17	56.32± 16.70	67.35± 25.10	92.54 88.08	6.20 9.33	1.26 2.59
March	3319.24± 158.48	2544.81± 208.72	3100.62± 140.64	2286.85± 236.24	178.62± 15.16	200.67± 12.10	40.00± 2.68	77.14± 16.00	93.41 89.08	5.38 7.89	1.21 3.03
April	2472.92± 168.80	2325.52± 296.28	2274.32± 128.35	2267.00± 180.62	146.15± 25.00	116.50± 12.00	52.45± 15.45	28.02± 16.28	91.97 93.79	5.91 5.01	2.12 1.20
May	2845.66± 433.10	1853.00± 191.67	2692.58± 374.60	2181.00± 268.00	126.72± 38.62	106.00± 15.68	26.36± 19.88	26.00± 39.42	94.62 92.88	4.45 5.72	0.93 1.40
June	1121.00± 184.69	1093.19± 48.38	987.64± 153.89	1721.00± 136.57	75.16± 24.00	98.72± 13.56	58.20± 6.80	6.85± 2.18	88.10 90.34	6.70 9.03	5.19 0.63
July	659.03± 80.20	710.35± 137.07	639.99± 72.10	987.62± 32.64	14.72± 5.62	20.68± 6.35	4.32± 2.48	5.42± 2.10	97.11 96.33	2.23 2.91	0.66 0.76
August	854.22± 49.49	761.04± 46.18	832.20± 32.68	684.25± 128.62	15.17± 12.00	37.10± 15.08	6.85± 4.81	8.32± 6.10	97.42 94.03	1.78 4.87	0.80 1.09
September	1025.52± 110.65	978.05± 80.09	936.54± 78.80	715.62± 25.00	12.16± 18.15	127.00± 5.62	16.82± 13.70	15.40± 16.35	91.32 85.44	7.04 12.99	1.64 1.57
October	1237.19± 68.83	1052.82± 35.12	1020.67± 53.61	835.65± 58.12	186.00± 2.54	168.15± 4.16	30.52± 12.68	16.40± 5.50	82.50 82.47	15.03 15.97	2.47 1.56
November	1225.09± 90.67	1240.33± 46.56	960.61± 58.74	868.27± 25.46	265.00± 28.67	262.00± 5.15	9.48± 3.26	10.12± 3.16	77.60 78.06	21.63 21.12	0.77 0.82
December	2471.62± 163.14	1598.21± 142.27	2172.00± 138.67	968.21± 38.25	326.08± 25.65	315.65± 12.17	84.00± 2.61	72.56± 2.10	87.88 75.71	8.72 19.75	3.40 4.54
Average	**2096.85± 257.14**	**1572.18± 144.61**	**1899.33± 228.06**	**1364.79± 121.17**	**158.30± 20.28**	**171.89± 11.25**	**39.21± 8.80**	**35.51± 12.19**	**90.21 87.03**	**7.88 11.00**	**1.97 1.91**

Sd±Standard Deviation

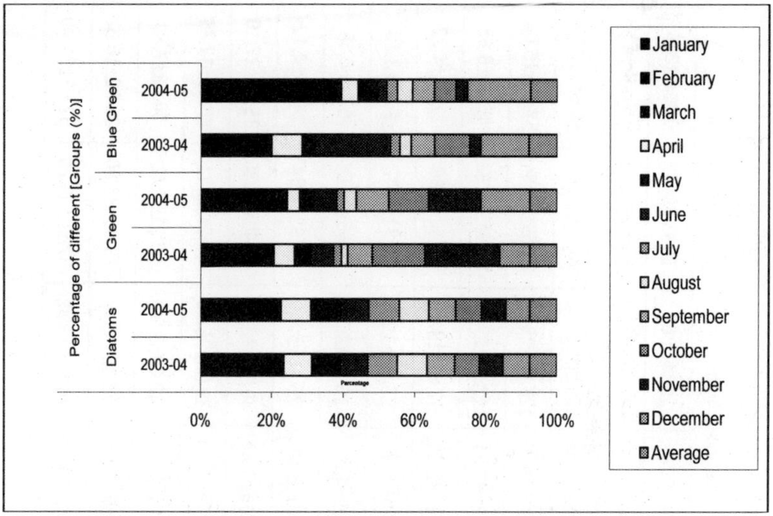

Fig. 4b.12: Graph showing percentage of different groups of phytoplankton at Sampling station C

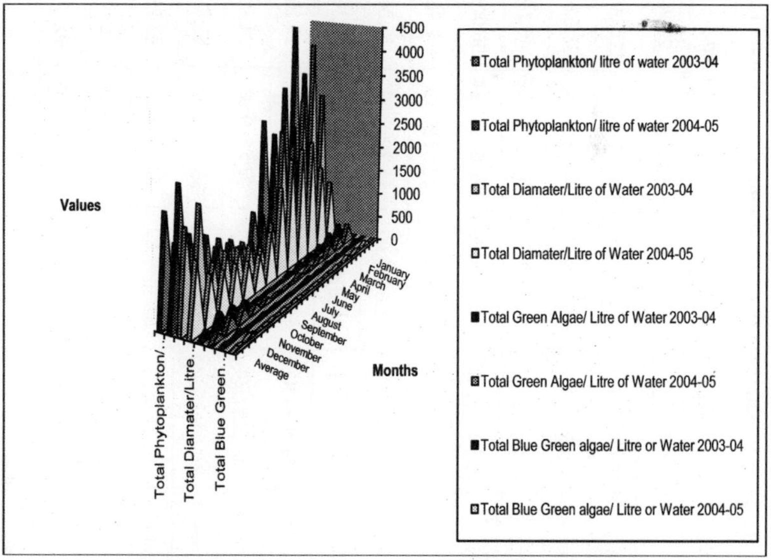

Fig. 4b.13: Graph showing values of different groups phytoplankton at Sampling station C

Table 4b.9: Number of Genera among Zooplankton in the River Ganga During 2003-05 at Sampling Station C

Months	*Keratella valga*		*Arcella*		*Ceriodaphnia*		*Filinia sp.*	
	2003-2004	2004-05	2003-2004	2004-2005	2003-2004	2004-2005	2003-2004	2004-2005
January	149.65±56.30	96.15±15.46	112.61±18.31	116.00±20.00	21.05±75.16	196.10±25.00	40.25±176.35	172.00±19.20
February	119.50±70.25	32.46±8.64	137.25±58.82	58.62±13.10	15.64±72.80	175.00±14.50	90.10±15.60	85.68±12.15
March	114.62±17.20	25.42±5.68	107.82±17.82	46.24±6.15	150.00±19.37	125.68±25.86	87.63±27.84	57.00±15.04
April	112.40±27.80	17.50±6.10	125.64±27.84	96.00±32.16	16.72±37.84	117.60±27.00	65.36±17.82	45.62±8.68
May	72.00±13.21	12.60±7.20	32.62±9.81	52.46±7.82	64.71±32.61	52.65±18.00	15.42±8.89	12.16±4.27
June	30.02±17.29	11.66±9.00	15.20±9.83	16.24±8.65	35.40±7.86	25.00±3.16	12.31±3.42	9.67±3.10
July	10.61±7.21	10.20±3.61	12.33±7.24	11.26±3.12	11.00±4.62	15.68±3.16	7.66±6.10	6.14±0.68
August	14.24±3.82	15.18±7.00	15.24±6.89	15.16±7.10	14.20±8.00	25.16±5.87	15.92±8.34	14.25±4.12
September	32.51±7.24	28.20±12.70	37.81±9.80	24.10±4.80	25.67±17.82	36.17±32.00	40.40±37.80	35.25±27.10
October	40.00±25.21	45.00±15.60	56.72±17.81	62.50±6.68	30.27±27.00	38.00±26.86	56.71±8.21	45.60±17.85
November	45.62±7.14	62.16±3.62	84.00±7.24	67.21±8.16	48.16±2.15	40.00±5.80	63.21±7.22	56.28±7.80
December	85.73±3.50	75.62±8.68	28.54±7.21	35.68±8.60	85.60±7.14	72.15±8.64	75.80±32.80	85.00±16.82
Average	68.91±21.35	36.01±8.61	63.82±16.55	50.12±10.53	43.20±26.03	76.60±16.32	47.56±29.20	52.05±11.40

Sd±Standard Deviation

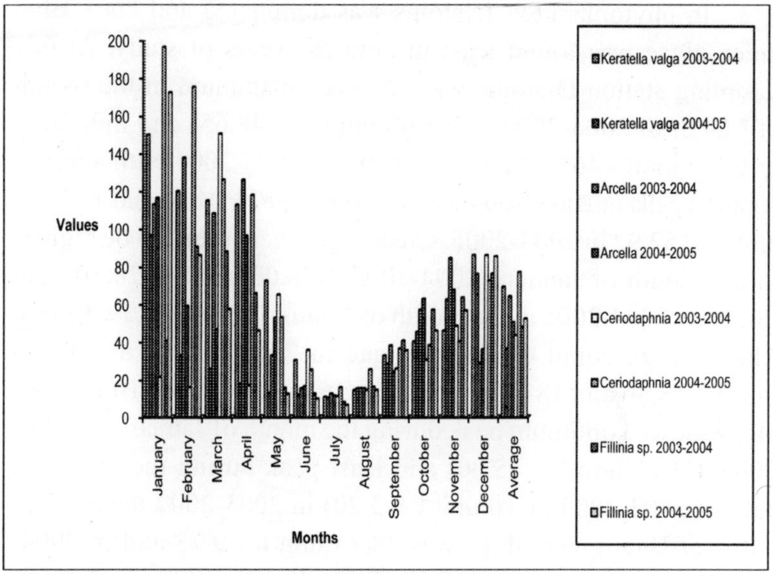

Fig. 4b.14: Graph showing values of different groups Zooplankton at Sampling station C

Sampling station D

(a) Plankton:

Quantitative analysis of plankton at Sampling station D revealed that the total Planktonic number per litre of water was maximum 3817.68 unit/l ± 477.20 in the month of January in the year 2004-2005 and minimum was 423.94 unit/l ± 209.82 in July month of the 2004-2005 (Table 4b.10-Table 4b.12 and Figure 4b.13- Figure 4b.16).

1. Phytoplankton :

Phytoplankton was noted lowest of 617.46 unit/l ± 32.57 in the month of July of 2003-2004 and highest of 4129.00 unit/l ± 291.60 in March month of 2004-2005 with the average cells number was 1375.10 unit/l ± 117.69 in 2003-2004 and 1999.68 unit/l ± 191.35 in 2004-2005 (Table-4b.10 and Figure 4b.13).

In phytoplankton Diatoms was dominated and class Blue green algae was found least in both the years of study. At this sampling station Diatoms was observed maximum in the month of March of 2004-2005 (3721.00 unit/l ± 248.65) and minimum in July month (462.10 unit/l ± 48.19) of 2003-2004 with average count of cells in 2003-2004 was 1230.09 unit/l ± 94.16 and 1775.10 unit/l ± 152.92 in 2004-2005. Green algae were found to be highest in the month of January 2004-2005 (378.00 unit/l ± 28.10) and lowest in 2004-2005 in the month of August (32.36 unit/l ± 15.00). The average count of Green algae in 2003-2004 was 116.28 unit/l ± 13.80 and 181.26 unit/l ± 26.16 in 2004-2005. Blue green algae showed optimum peak during the month of January in 2004-2005 (82.12 unit/l ± 15.00) and least peak during the month of July of 2003-2004 (1.16 unit/l ± 3.20) in 2003-2004 the average value of Blue green algae was 28.73 unit/l ± 9.73 and in 2004-2005 it was 43.32 unit/l ± 12.27 (Table 4b.11 and Figure 4b.14 to Figure 4b.15).

2. Zooplankton :

Zooplankton was observed minimum in August of 2003-2004 (49.11 unit/l ± 20.81) and maximum in 2004-2005 for the month of January (397.64 unit/l ± 116.78). And the average value of 257.69 unit/l ± 87.93 in 2003-2004 and in 2004-2005 the average was 213.40 unit/l ± 42.00. (Table 4b.10 and Table 4b.12) and (Figure 4b.13 and Figure 4b.16).

At this spot *Keretala valga* was maximum in the month of January in 2004-2005 (162.00 unit/l ± 24.32) and minimum was found 12.37 unit/l ± 6.24 in July month. (2004-2005). *Ceriodapnia* was found to be highest in the month of April 2004-2005 (98.62 unit/l ± 8.67) and lowest (5.14 unit/l ± 3.00) in the month of August of the year 2003-2004. *Arcella* was observed in their optimum peak during 2004-2005 November (78.26 unit/l ± 3.26) and in lowest peak during the month of August (2.00 unit/l ± 0.00) in the year 2003-2004. *Crustacea* was maximum (92.35 unit/l ± 35.12) in 2003-2004 March and minimum was found in the month of August 2003-2004 (11.36 unit/l ± 5.10). *Fillinia* sp. was observed in their optimum peak during 2004-2005 January (112.00 unit/l ± 64.00) and in lowest peak during the month of August (4.50unit/l ± 4.20) in the year 2003-2004 (Table 4b.12 and Figure. 4b.16).

The average value revealed that *Keretala valga* was dominating in the zooplankton, 2003-2004 (58.65unit/l ± 11.77) followed by *Ceriodapnia* (42.09 unit/l ± 9.22), *Crustacea* (40.65 unit/l ± 10.60), *Fillinia* sp. (29.27 unit/l ± 7.05), Arcella (9.36 unit/l ± 1.95). In second year 2004-2005 *Keretala valga* (78.87 unit/l ± 14.51) found in strength followed by Arcella (48.68 unit/l ± 7.51), *Ceriodaphnia valga* (48.58 unit/l ± 8.08) and *Fillinia* sp. (37.28 unit/l ± 11.89) (Table 4b.12 and Figure 4b.16).

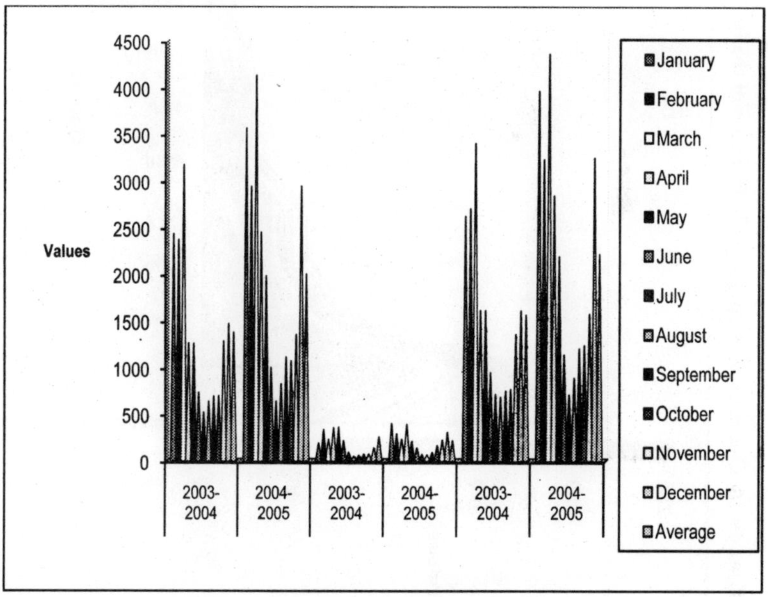

Fig. 4b.16: Graph showing monthly fluctuation of total plankton at Sampling station D

Table 4b.10: Quantitative Analysis of the Plankton of the River Ganga During 2003-2005 at Sampling Station D

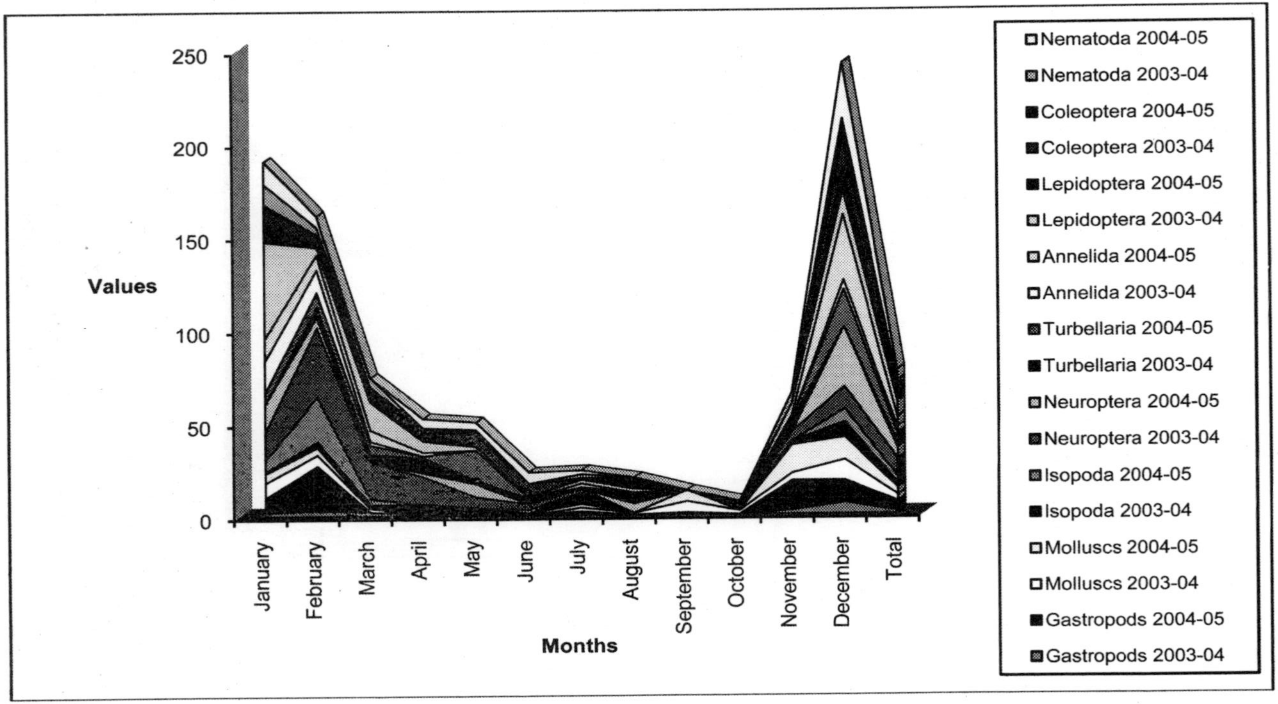

Sd±Standard Deviation

Table 4b.11: Number and Percentage of different Group among the Phytoplankton of the River Ganga During 2003-05 at Sampling Station D

Months	Total Phytoplankton/ litre of water		Total Diatoms /Litre of Water		Total Green Algae/ Litre of Water		Total Blue Green algae/ Litre or Water		Percentage of different [Groups (%)]		
	2003-2004	**2004-2005**	**2003-2004**	**2004-05**	**2003-04**	**2004-05**	2003-04	2004-05	**Diatoms**	**Green**	**Blue Green**
January	2431.02 ±154.34	3560.70 ±189.34	2101.00 ±126.24	3100.58 ±146.24	268.00 ±18.10	378.00 ±28.10	62.02 ±10.00	82.12 ±15.00	86.42±87.08	11.02±10.62	2.55±2.31
February	2369.50 ±156.16	2939.50 ±166.16	2205.25 ±116.00	2765.25 ±126.00	106.25 ±24.14	146.25 ±24.14	58.00 ±16.02	28.00 ±16.02	93.07±94.07	44.98.48±	2.45±0.95
March	3169.80 ±171.60	4129.00 ±291.60	3021.00 ±148.65	3721.00 ±248.65	102.80 ±6.15	362.00 ±26.15	46.00 ±16.80	46.00 ±16.80	95.31±90.12	3.24±8.77	1.45±1.11
April	1259.84 ±261.64	2449.84 ±375.24	1108.52 ±241.62	2108.52 ±344.62	102.68 ±5.00	282.68 ±15.00	48.64 ±15.02	58.64 ±15.62	87.99±86.07	8.15±11.54	3.86±2.39
May	1257.67 ±147.62	1979.66 ±176.82	1161.04 ±105.62	1861.04 ±125.62	65.82 ±32.00	85.81 ±32.00	30.81 ±10.00	32.81 ±19.20	92.32±94.01	5.23±4.33	2.45±1.66
June	728.62 ±75.26	999.86 ±55.36	656.00 ±55.00	856.24 ±25.00	68.00 ±15.10	98.00 ±15.10	4.62 ±5.16	45.62 ±15.26	90.03±85.64	9.33±9.80	0.63±4.56
July	517.46 ±62.57	635.62 ±43.47	462.10 ±48.19	562.20 ±28.19	54.20 ±11.18	64.26 ±12.18	1.16 ±3.20	9.16 ±3.10	89.30±88.45	10.47±10.11	0.22±1.44
August	631.45 ±53.25	821.71 ±32.93	586.05 ±32.00	786.25 ±12.68	42.30 ±15.00	32.36 ±15.00	3.10 ±6.25	3.10 ±5.25	92.81±95.68	6.70±3.94	0.49±0.38
September	688.62 ±41.71	1112.62 ±74.06	615.00 ±12.01	985.00 ±32.76	65.62 ±15.08	85.62 ±25.68	8.00 ±14.62	42.00 ±15.62	89.31±88.53	9.53±7.70	1.16±3.77
October	693.00 ±30.66	1073.00 ±89.06	565.25 ±12.42	865.25 ±12.42	105.60 ±6.16	125.60 ±62.56	22.15 ±12.08	82.15 ±14.08	81.57±80.64	15.24±11.71	3.20±7.66
November	1282.88 ±128.82	1352.88 ±451.42	1065.62 ±120.06	1075.62 ±420.36	212.00 ±5.12	252.00 ±25.42	5.26 ±3.64	25.26 ±5.64	83.06±79.51	16.53±18.63	0.41±1.87
December	1471.32 ±128.70	2941.72 ±350.70	1214.20 ±112.12	2614.20 ±312.52	202.12 ±12.56	262.52 ±32.56	55.00 ±4.02	65.00 ±5.62	82.52±88.87	13.74±8.92	3.74±2.21
Average	**1375.10** ±117.69	**1999.68** ±191.35	**1230.09** ±94.16	**1775.10** ±152.92	**116.28** ±13.80	**181.26** ±26.16	**28.73** ±9.73	**43.32** ±12.27	**88.64**±88.22	**9.47**±9.25	**1.88**±2.53

Sd±Standard Deviation

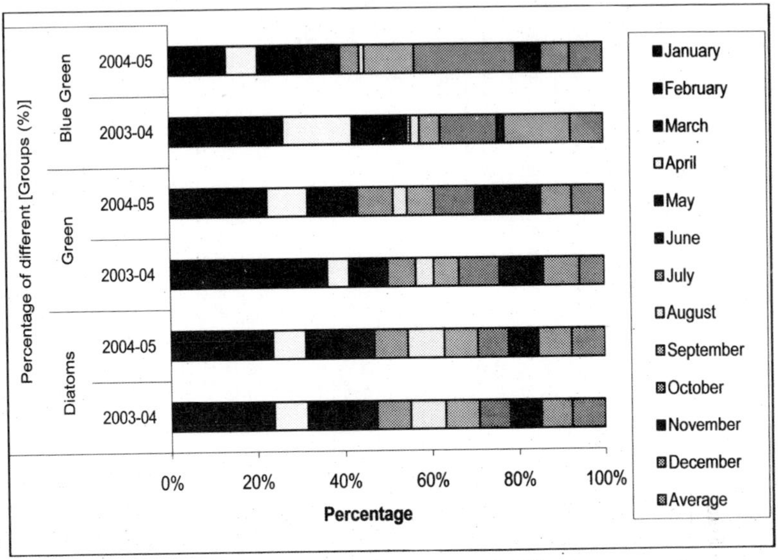

Fig. 4b.17: Graph showing percentage of different groups of phytoplankton at Sampling station D

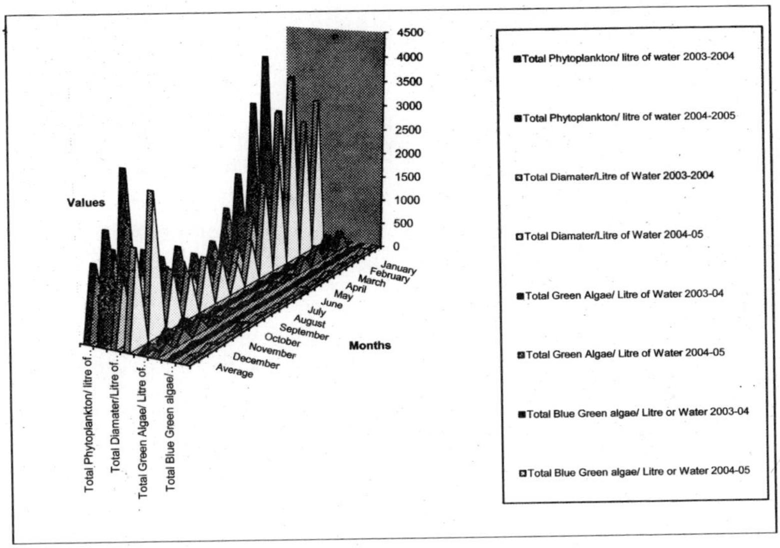

Fig. 4b.18: Graph showing values of different groups phytoplankton at Sampling station D

Table 4b.12: Number of Different Genera among Zooplankton in the River Gnaga during 2003-2005 at Sampling Station D

Months	*Keratella valga*		*Ceriodaphnia*		*Arcella*		*Crustacea*		*Fillinia Sp.*	
	2003-2004	2004-2005	200-2004	2004-2005	2003-2004	2004-2005	2003-2004	2004-2005	2003-2004	2004-2005
January	145.00±14.26	162.00±24.32	68.20±12.00	65.37±15.00	12.10±3.10	58.27±13.46	55.00±10.00	-	54.20±11.16	112.00±64.00
February	62.14±12.50	136.25±68.15	50.10±5.24	25.60±8.24	16.20±2.16	48.62±0.26	86.00±12.21	-	15.10±5.11	78.00±18.46
March	104.32±36.20	142.00±16.52	66.00±14.21	15.62±6.16	22.00±5.00	36.00±6.21	92.35±35.12	-	68.00±22.15	35.62±7.81
April	106.05±15.04	136.25±17.21	86.00±15.10	98.62±8.67	15.01±2.20	135.65±8.27	68.00±16.14	-	82.56±5.86	20.00±6.20
May	62.05±12.30	62.14±8.68	65.00±10.14	72.46±8.21	10.65±1.10	56.00±6.08	52.02±12.24	-	25.00±6.20	15.24±3.16
June	42.24±15.20	52.56±8.67	25.62±2.60	52.25±2.40	3.10±1.04	24.00±7.80	10.00±2.64	-	11.10±6.20	8.10±6.00
July	16.08±8.59	12.37±6.24	16.81±2.50	35.40±1.68	5.00±2.00	12.72±8.16	6.12±5.62	-	5.10±2.10	7.81±1.80
August	35.14±6.20	15.13±2.16	5.14±3.00	26.66±7.86	2.00±0.00	16.32±3.25	11.36±5.10	-	4.50±4.20	6.00±1.06
September	25.00±5.10	18.21±3.62	12.00±12.10	30.50±8.24	6.60±3.50	25.40±6.68	14.61±3.76	-	12.56±3.40	16.71±2.46
October	20.10±3.16	56.27±8.62	12.00±11.40	45.16±6.08	12.30±2.12	36.72±8.10	14.20±2.10	-	15.06±2.16	25.36±8.40
November	35.52±2.50	67.00±2.06	52.10±11.30	40.00±5.80	2.15±0.21	78.26±3.26	26.00±11.00	-	22.00±5.00	36.27±5.75
December	50.10±10.15	86.24±7.86	46.12±11.00	75.26±18.62	5.16±1.00	56.15±18.62	52.10±11.26	-	36.00±11.10	86.24±17.62
Average	58.65±11.77	78.87±14.51	42.09±9.22	48.58±8.08	9.36±1.95	48.68±7.51	40.65±10.60	-	29.27±7.05	37.28±11.89

Sd±Standard Deviation

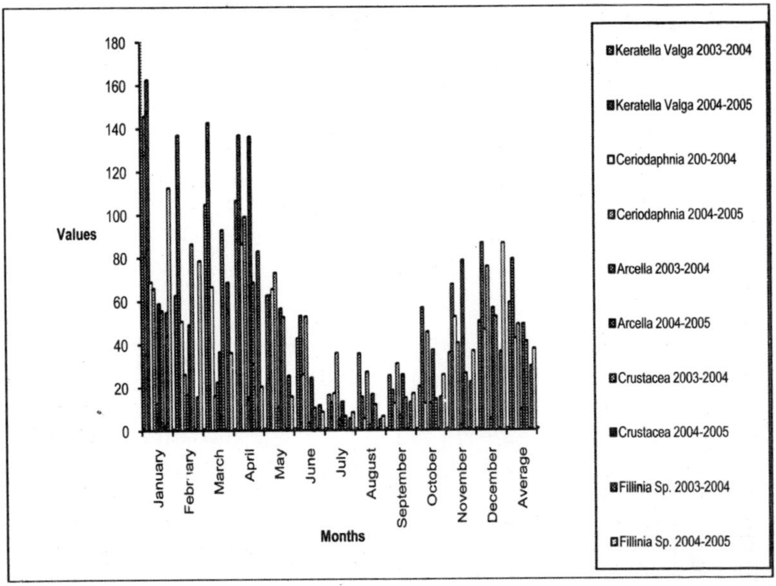

Fig. 4b.19: Graph showing values of different groups Zooplankton at Sampling station D

Sampling station E

(a) Plankton :

Quantitative analysis of plankton at Sampling station E revealed that the total planktonic number per litre of water was maximum 3817.68 unit/l ± 477.20 in the month of January in the year 2004-2005 and minimum was 423.94 unit/l ± 209.82 in July month of the 2004-2005 (Table 4b.13-Table 4b.15 and Figure 4b.17- Figure 4b.20).

1. Phytoplankton :

Phytoplankton was noted lowest of 518.90 unit/l ± 110.56 in the month of July of 2004-2005 and highest of 8933.93 unit/l ± 300.62 in January month of 2003-2004, with the average cells number was 3454.49 unit/l ± 342.80 in 2003-2004 and 1639.03 unit/l ± 161.12 in 2004-2005 (Table-4b.13 and Figure 4b.17).

In phytoplankton Diatoms was dominated and class Blue green algae was found least in both the years of study. At this sampling station Diatoms was observed maximum in the month of January of 2003-2004 (8261.80 unit/l ± 178.26) and minimum in July month (324.36 unit/l ± 82.50) of 2004-2005 with average count of cells in 2003-2004 was 3168.41 unit/l ± 294.54 and 1401.63 unit/l ± 127.88 in 2004-2005. Green algae were found to be highest in the month of January 2003-2004 (585.65 unit/l ± 98.20) and lowest in 2004-2005 in the month of August (28.10 unit/l ± 5.62). The average count of Green algae in 2003-2004 was 239.59 unit/l ± 34.93 and 180.35 unit/l ± 18.37 in 2004-2005. Blue green algae showed optimum peak during the month of February in 2004-2005 (167.85 unit/l ± 12.37) and least peak during the month of July of 2004-2005 (2.72 unit/l ± 18.00) in 2003-2004 the average value of Blue green algae was 46.49 unit/l ± 13.33 and in 2004-2005 it was 57.06 unit/l ± 14.86 (Table 4b.14 and Figure 4b.18 to Figure 4b.19).

2. Zooplankton :

Zooplankton was observed minimum in July of 2004-2005 (19.00 unit/l ± 17.46) and maximum in 2004-2005 for the month of January (519.98 unit/l ± 100.22) and the average value of 219.60 unit/l ± 51.87 in 2003-2004 and in 2004-2005 the average was 214.91 unit/l ± 38.57 (Table 4b.13 and Table 4b.15) and (Figure 4b.17 and Figure 4b.20).

At this spot *Keretala valga* was maximum in the month of January in 2004-2005 (162.35 unit/l ± 26.47) and minimum was found 2.65-unit/l ± 1.26 in July month. (2004-2005). *Ceriodapnia* was found to be highest in the month of January 2004-2005 (162.24 unit/l ± 26.32) and lowest (5.25 unit/l ± 1.80) in the month of July of the same year (2004-2005). *Arcella* was observed in their optimum peak during 2004-2005 January (124.35 unit/l ± 6.40) and in lowest peak during the month of August (1.08 unit/l ± 1.24) in the year 2003-2004. Crustacea was maximum (115.26 unit/l ± 28.30) in 2003-2004 February and minimum was found in the month of July 2003-2004 (3.24 unit/l ± 1.67). *Fillinia* sp. was observed in their optimum peak during 2003-2004 March (118.05 unit/l ± 75.80) and in lowest peak during the month of July (2.10 unit/l ± 0.18) in the year 2003-2004 (Table 4b.15 and Fig. 4b.20).

The average value revealed that *Ceriodapnia* was dominating in the zooplankton, 2003-2004 (76.08unit/l ± 27.40) followed by *Keretala valga* (61.61 unit/l ± 19.62), *Crustacea* (58.25 unit/l ± 17.68), *Fillinia* sp. (46.36 unit/l ± 18.46), *Arcella* (15.39 unit/l ± 4.78). In second year 2004-2005 *Ceriodapnia* (76.08 unit/l ± 27.40) found in strength followed by *Arcella* (58.25 unit/l ± 17.68), *Keretala valga* (58.12 unit/l ± 13.60) and *Fillinia* sp. (46.81 unit/l ± 11.78) (Table 4b.15 and Fig. 4b.20).

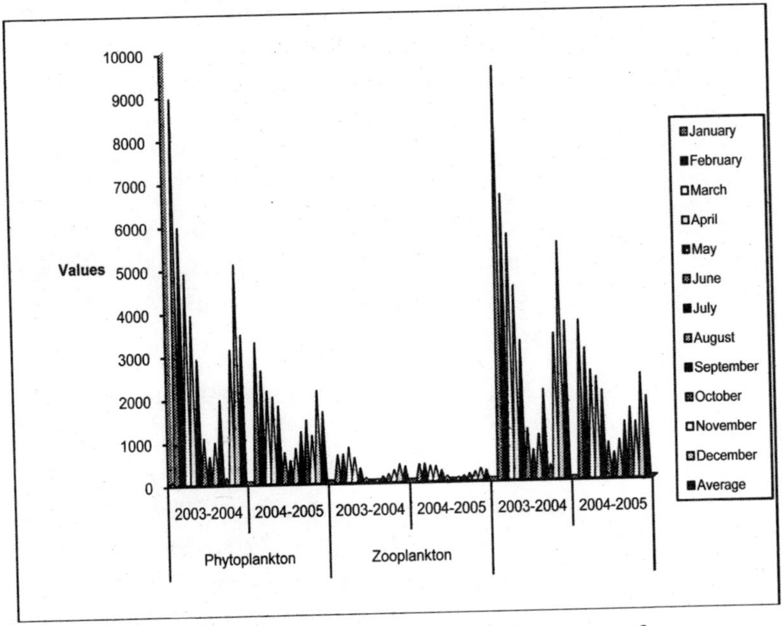

Fig. 4b.21: Graph showing monthly fluctuation of total plankton at Sampling station E

Table 4b.13: Quantitative Analysis of the Plankton of the River Ganga During 2003-2005 at Sampling Station E

Months	Number of Plankton per liter of water				Total Plankton	
	Phytoplankton		Zooplankton			
	2003-2004	**2004-2005**	**2003-2004**	**2004-2005**	**2003-2004**	**2004-2005**
January	8933.93±300.62	3272.40±423.21	627.30±162.82	376.80±147.70	9561.23±463.44	3649.2±570.91
February	5952.35±288.42	2609.36±146.59	647.65±186.00	391.02±126.47	6600±474.42	3000.38±273.06
March	4876.00±833.99	2138.29±174.83	799.53±240.81	343.72±77.73	5675.53±1074.8	2482.01±252.56
April	3911.16±871.88	1991.89±168.39	558.71±67.79	333.01±84.60	4469.87±939.67	2324.9±252.99
May	2895.00±130.99	1783.98±74.81	311.74±190.71	230.30±49.20	3206.74±321.70	2014.28±324.01
June	1065.55±198.37	704.84±69.27	89.08±31.12	103.57±34.24	1154.63±229.49	808.41±103.51
July	635.44±56.77	518.90±110.56	34.21±11.55	56.42±13.82	669.65±68.32	575.32±124.28
August	964.56±195.36	795.84±23.53	54.75±12.51	66.84±16.61	1019.31±207.87	862.68±40.14
September	1940.68±209.44	1184.97±114.83	115.92±25.25	99.99±39.99	2056.6±234.69	1284.96±154.82
October	140.57±183.67	1453.86±118.63	166.60±33.01	150.31±35.78	307.17±216.68	1604.17±154.41
November	3099.00±472.72	1090.04±96.49	257.07±43.90	189.53±36.70	3356.07±516.62	1279.57±133.19
December	5079.60±371.35	2124.04±212.124	389.27±130.83	262.22±43.15	5468.87±502.18	2386.26±255.274
Average	**3454.49±342.80**	**1639.03±161.12**	**337.65±94.69**	**216.98±58.83**	**3628.80±437.49**	**1856.01±219.01**

Sd±Standard Deviation

Table 4b.14: Number and Percentage of different Group among the Phytoplankton of the River Ganga During 2003-05 at Sampling Station E

Months	Total Phytoplankton/ litre of water		Total Diatoms/Litre of Water		Total Green Algae/ Litre of Water		Total Blue Green algae/ Litre or Water		Percentage of different (Groups (%)		
	2003-2004	2004-2005	2003-2004	2004-2005	2003-2004	2004-2005	2003-2004	2004-2005	Diatoms	Green	Blue Green
January	8933.93± 300.62	3272.40± 423.21	8261.80± 178.26	2804.62± 362.80	585.65± 98.20	362.00± 25.17	86.48± 24.16	105.78± 35.24	92.48±85.71	6.56±11.06	0.97±3.23
February	5952.35± 288.42	2609.36± 146.59	5682.00± 228.62	2176.35± 118.54	246.35± 42.48	265.16± 15.68	24.00± 17.32	167.85± 12.37	95.46±83.41	4.14±10.16	0.40±6.43
March	4876.00± 833.99	2138.29± 174.83	4326.00± 782.15	1895.67± 132.00	510.00± 36.74	200.62± 27.16	40.00± 15.10	42.00± 15.67	88.72±88.65	10.46±9.38	0.82±1.96
April	3911.16± 871.88	1991.89± 168.39	3468.78± 844.16	1765± 130.62	386.98± 12.20	168.25± 15.65	55.40± 15.52	58.64± 22.12	88.69±88.61	9.89±8.45	1.42±2.94
May	2895.00± 130.99	1783.98± 274.81	2782.00± 82.63	1625.54± 246.20	75.20± 28.64	132.10± 16.21	37.80± 19.72	26.34± 12.40	96.10±91.12	2.60±7.40	1.31±1.48
June	1065.55± 198.37	704.84± 69.27	927.30± 175.35	624.70± 50.10	97.80± 12.32	74.14± 8.17	40.45± 10.70	6.00± 11.00	87.03±88.63	9.18±10.52	3.80±0.85
July	635.44± 56.77	518.90± 110.56	580.40± 39.45	324.36± 82.50	46.54± 11.20	191.82± 10.06	8.50± 6.12	2.72± 18.00	91.34±62.51	7.32±36.97	1.34±0.52
August	964.56± 195.36	795.84± 23.53	1625.85± 172.06	762.37± 12.46	35.67± 18.10	28.10± 5.62	3.04± 5.20	5.37± 5.45	97.67±95.79	2.14±3.53	0.18±0.67
September	1940.68± 209.44	1184.97± 114.83	1750.00± 132.78	1062.71± 86.37	125.68± 58.26	95.62± 18.75	65.00± 18.40	26.64± 9.71	90.17±89.68	6.48±8.07	3.35±2.25
October	140.57± 183.67	1453.86± 118.63	1192.89± 168.37	1232.54± 78.24	135.42± 2.00	162.00± 25.15	72.26± 13.30	59.32± 15.24	85.17±84.78	9.67±11.14	5.16±4.08
November	3099.00± 472.72	1090.04± 96.49	2799.56± 430.14	800± 56.37	256.94± 36.40	268.72± 35.27	42.50± 6.18	21.32± 4.85	90.34±73.39	8.29±24.65	1.37±1.96
December	5079.60± 371.35	2124.04± 212.124	4624.30± 300.52	1745.68± 178.36	372.80± 62.56	215.62± 17.56	82.50± 8.27	162.74± 16.32	91.04±82.19	7.34±10.15	**1.62**±7.66
Average	**3454.49± 342.80**	**1639.03**± 161.12	**3168.41± 294.54**	**1401.63**± **127.88**	**239.59**± 34.93	**180.35**± 18.37	**46.49**± 13.33	**57.06**± 14.86	**91.18**±84.54	**7.01**±12.62	**1.81**±2.84

Sd±Standard Deviation

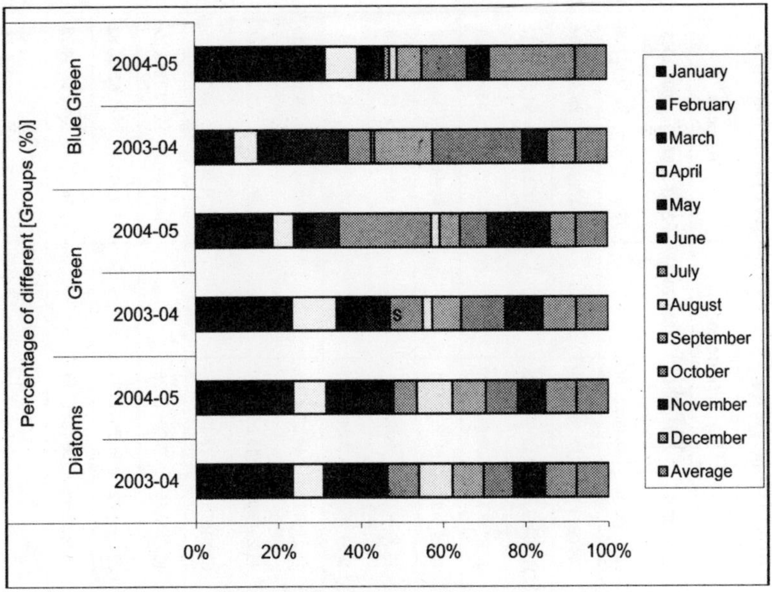

Fig. 4b.22: Graph showing percentage of different groups of phytoplankton at Sampling station E

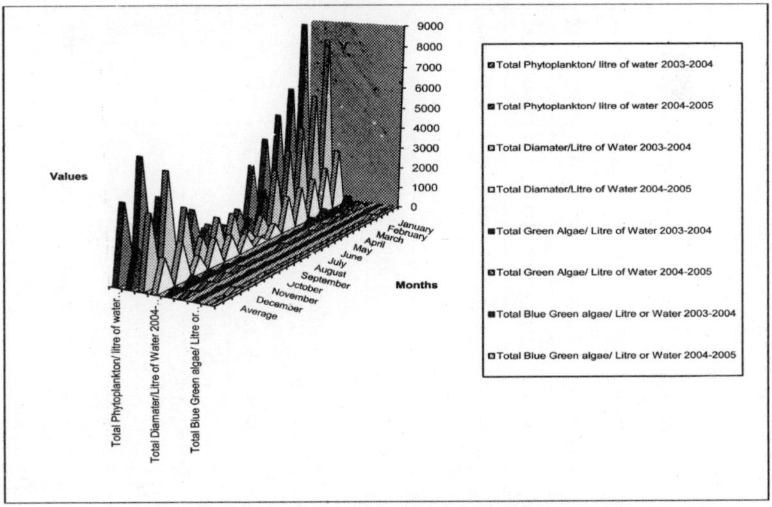

Fig. 4b.23: Graph showing values of different groups phytoplankton at Sampling station E

Table 4b.15: Number of Different Genera among Zooplankton in the River Ganga during 2003-2005 at Sampling Station E

Months	*Keratella valga*		*Ceriodaphnia*		*Arcella*		*Crustacea*		*Fillinia Sp.*	
	2003-2004	**2004-2005**	**2003-2004**	**2004-2005**	**2003-2004**	**2004-2005**	**2003-2004**	**2004-2005**	**2003-2004**	**2004-2005**
January	116.00±56.31	162.35±26.47	116.60±28.26	162.24±26.32	17.56±8.64	124.35±6.40	106.00±37.84	-	78.62±25.62	96.35±42.09
February	76.84±29.82	108.60±36.87	106.92±56.81	106.20±12.50	37.70±8.12	104.00±9.56	115.26±28.30	-	54.17±18.60	86.87±16.24
March	112.67±39.82	112.10±18.26	112.84±88.92	185.67±57.82	45.00±5.04	117.63±7.08	115.00±35.26	-	118.05±75.80	94.27±15.12
April	105.00±16.08	40.32±5.42	192.30±16.24	90.00±13.62	25.62±8.20	36.87±8.62	100.62±27.80	-	84.35±6.42	15.64±8.27
May	97.42±18.86	56.14±16.54	53.00±30.24	96.52±24.00	16.00±5.20	78.24±6.12	55.65±27.84	-	57.48±50.60	34.26±9.32
June	46.65±23.50	12.27±2.62	17.00±4.90	17.16±8.27	3.25±1.86	25.30±6.42	15.60±7.23	-	17.86±6.30	15.26±6.15
July	16.70±7.40	2.65±1.26	5.85±4.62	5.25±1.80	1.80±0.02	1.54±6.10	3.24±1.67	-	2.10±0.18	6.17±3.26
August	20.86±6.36	7.00±2.37	8.70±7.00	6.35±4.20	**1.08**±1.24	10.21±4.28	14.00±6.10	-	3.00±1.56	8.10±6.21
September	25.62±8.82	12.54±7.30	52.50±17.24	27.05±16.30	**6.70**±3.25	43.10±5.00	15.26±7.28	-	20.00±8.62	25.26±8.37
October	20.35±4.29	42.80±19.56	65.00±13.86	37.70±8.00	16.20±8.24	63.10±7.14	28.12±4.16	-	32.71±8.84	36.15±4.25
November	37.49±8.50	43.00±7.86	85.02±23.46	55.55±6.25	4.32±1.50	67.21±8.56	35.00±15.28	-	36.00±8.00	57.78±5.20
December	63.70±15.70	97.64±18.72	97.22±37.25	87.20±8.68	9.40±6.00	35.26±8.62	95.28±13.36	-	52.00±11.00	85.62±16.82
Average	61.61±19.62	58.12±13.60	76.08±27.40	73.07±15.65	15.39±4.78	58.90±6.99	58.25±17.68	-	**46.36**±18.46	46.81±11.78

Sd±Standard Deviation

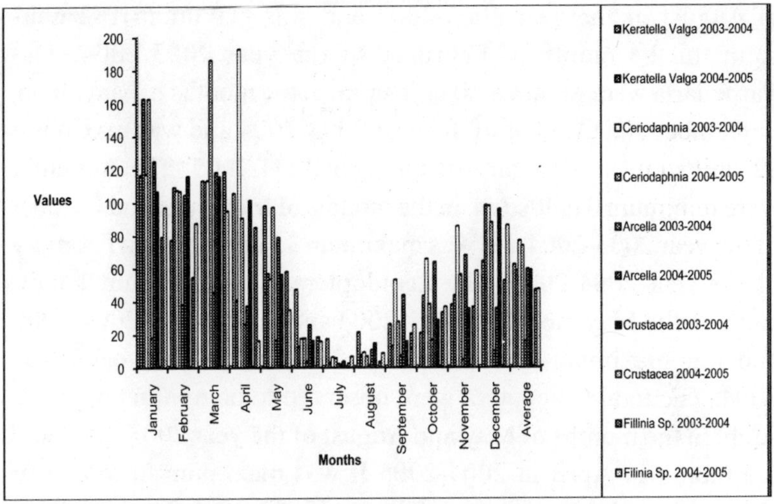

Fig. 4b.24: Graph showing values of different groups Zooplankton at Sampling station E

22. Micro Benthicfauna :

Sampling station A

The comparative observation at sampling station A showed that the Neuroptera shown the dominancy over the other benthicfauna with the average value of 48.25 unit/sq.m of the year 2003-2004 and the Molluscs with the minimum average value of 1.33 unit/sq.m of the year 2003-2004 (Table 4b.16 and figure 4b.21).

The Gastropods were observed to be minimum 1 unit/sq.m. in the months of May, June, and September of the year 2003-2004 and was maximum 17 unit/sq.m. in the month of December of the year 2004-2005.The Molluscs were minimum Nil found in the months of May, June July September and October of the year 2003-2004 and was maximum 6 unit/sq.m. in the months of November and December of the year 2004-2005. The Isopoda were nil in the months May, June, July and August of the year 2003-2004 and was maximum 18 unit/sq.m. in December of the year 2003-2004. The Neuroptera found to be minimum 2 unit/sq.m. in the month of August of the year 2003-2004 and was maximum 182 unit/sq.m. in the month of February of the year 2003-2004. The Turbellaria were minimum 0 unit/sq.m. in the months of May, June, September and October of the year 2003-2004 and was maximum 32 unit/sq.m. in February of the year 2004-2005. The Annelida were minimum 0 unit/sq.m. in the months of May, June and October of the year 2003-2004 and was maximum 35 unit/sq.m. in February of the year 2004-2005. The Lepidoptera were minimum 1 unit/sq.m. in the May month of 2003-2004 and in March of 2004-2005 and was maximum 53 unit/sq.m. in February of the year 2003-2004. The total Coleoptera were observed to be minimum 1 unit/sq.m. in the months of May and August of the year 2003-2004 and the month of April in 2004-2005 it was maximum in month of December with 18 unit/sq.m. of the year 2004-2005. The Nematoda were minimum 0 unit/sq.m. in the months of April, May, June, September, October and November of the year 2003-2004 and maximum 17 unit/sq.m. in the month of December of the year 2004-2005.

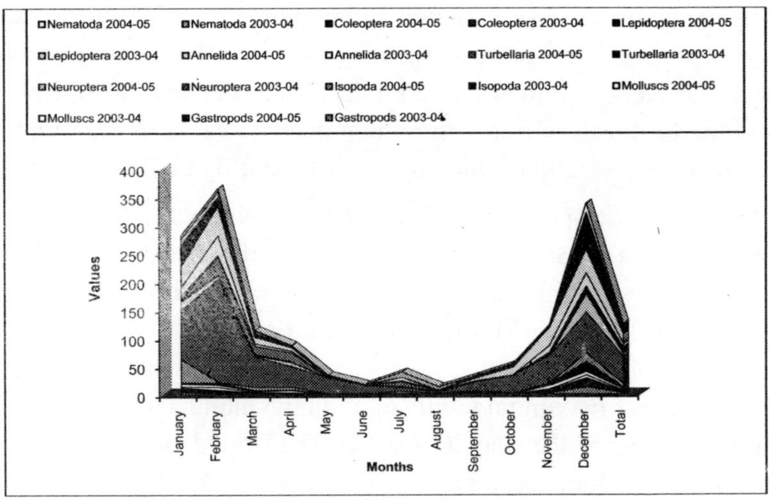

Fig. 4b.5: Graph showing values of different groups of macro benthos at Sampling station A

Table 4b.16: Benthicfauna in the river Ganga during 2003-2005 at Sampling Station-A

Months	*Gastropods*		*Molluscs*		*Isopoda*		*Neuroptera*		*Turbellaria*		*Annelida*		*Lepidoptera*		*Coleoptera*		Nematoda	
	2003-04	2004-05	2003-04	2004-05	2003-04	2004-05	2003-04	2004-05	2003-04	2004-05	2003-04	2004-05	2003-04	2004-05	2003-04	2004-05	2003-04	2004-05
January	4	8	2	5	2	42	86	15	5	-	5	15	37	15	8	14	12	6
February	2	6	5	4	4	1	182	6	2	32	2	35	53	5	4	6	11	5
March	3	-	1	2	1	6	54	2	2	15	4	-	12	1	3	2	4	2
April	2	-	1	4	1	5	37	7	1	16	3	-	6	-	2	1	0	1
May	1	-	0	-	0	-	28	-	0	-	0	-	2	2	1	-	0	-
June	1	-	0	-	0	-	15	-	0	-	0	-	1	-	-	-	0	-
July	-	2	0	3	0	-	7	-	2	8	2	6	-	-	-	-	10	-
August	-	-	1	1	0	-	2	-	1	-	1	-	-	-	1	-	6	-
September	1	-	0	2	1	2	16	-	0	-	2	-	5	2	2	-	0	-
October	-	-	0	-	2	-	32	-	0	-	0	-	14	-	4	2	0	-
November	2	6	1	6	4	1	48	-	5	-	1	12	32	-	3	-	0	2
December	12	17	5	6	18	15	72	32	11	-	2	24	42	35	6	18	6	17
Total	**2.33**	**3.25**	**1.33**	**2.75**	**2.75**	**6.00**	**48.25**	**5.17**	**2.42**	**5.92**	**1.83**	**7.67**	**17.00**	**5.00**	**2.83**	**3.58**	**4.08**	**2.75**

Sd±Standard Devi tion

Sampling station B

The comparative observation at sampling station B showed that the Neuroptera shown the dominancy over the other benthicfauna with the average value of 9.58 unit/sq.m. of the year 2003-2004 and the Gastropods with the minimum average value of 1.42 unit/sq.m. of the year 2003-2004 (Table 4b.17 and figure 4b.22).

The Gastropods were observed to be minimum 0 unit/sq.m. in the months of May, June, July, August and October of the year 2003-2004 and was maximum 24 unit/sq.m. in the month of February of the year 2004-2005.The Molluscs were minimum 1 unit/sq.m. found in the month of August in 2003-2004 and month of May and June of the year 2004-2005. The maximum of 15 unit/sq.m. were observed in the month of November of the year 2004-2005. The Isopoda were lowest 1 unit/sq.m. in the month of January of the year 2003-2004 and was maximum 24 unit/sq.m. in February of the year 2004-2005. The Neuroptera found to be minimum 2 unit/sq.m. in the month of July of the year 2003-2004 and month of April in the year 2004-2005 and was maximum 37 unit/sq.m. in the month of February of the year 2003-2004. The Turbellaria were minimum 1 unit/sq.m. in the month of July of the year 2003-2004 and month of April and November of the year 2004-2005. The value was maximum of 16 unit/sq.m. in May of the year 2004-2005. The Annelida were minimum 2 unit/sq.m. in the month of March and July of the year 2003-2004 and in the month of April and November of the year 2004-2005. It was maximized in December of the year 2004-2005 (35 unit/sq.m.). The Lepidoptera were minimum 1 unit/sq.m. in the June month of 2003-2004 and was maximum 52 unit/sq.m. in January of the year 2003-2004. The total Coleoptera were observed to be minimum 1 unit/sq.m. in the month of July of the year 2004-2005 it was maximized in month of December with 17 unit/sq.m. of the year 2003-2004. The Nematoda were minimum 1 unit/sq.m. in the month of October of the year 2004-2005 and maximum 30 unit/sq.m. in the month of December of the year 2004-2005.

Table 4b.17: Benthicfauna in the river Ganga during 2003-2005 at Sampling Station-B

Months	*Gastropods*		*Molluscs*		*Isopoda*		*Neuroptera*		*Turbellaria*		*Annelida*		*Lepidoptera*		*Coleoptera*		*Nematoda*	
	2003-04	2004-05	2003-04	2004-05	2003-04	2004-05	2003-04	2004-05	2003-04	2004-05	2003-04	2004-05	2003-04	2004-05	2003-04	2004-05	2003-04	2004-05
January	2	8	8	5	1	6	15	15	-	6	17	12	52	10	8	2	11	12
February	3	24	6	4	3	24	37	6	5	8	12	8	4	2	5	-	4	6
March	2	-	2	-	2	3	20	-	4	5	2	6	15	6	2	-	5	-
April	1	-	-	-	5	-	15	2	8	1	-	2	6	5	-	2	-	5
May	0	-	-	1	3	-	6	7	2	16	-	-	2	-	6	3	-	4
June	0	-	-	1	-	2	3	-	3	-	-	-	1	-	7	-	-	6
July	0	2	2	2	6	-	2	-	1	-	2	-	2	-	-	1	2	2
August	0	-	1	-	-	-	-	5	6	-	-	-	2	-	-	6	-	-
September	1	-	6	6	-	-	-	-	-	-	-	-	-	-	-	-	-	-
October	0	-	2	-	-	-	-	-	-	2	-	-	-	2	-	-	-	1
November	2	16	4	15	2	-	5	-	6	1	-	2	6	-	-	-	6	-
December	6	12	11	12	8	7	12	32	-	20	5	35	13	12	17	5	5	30
Total	**1.42**	**5.17**	**3.50**	**3.83**	**2.50**	**3.50**	**9.58**	**5.58**	**2.92**	**4.92**	**3.17**	**5.42**	**8.58**	**3.08**	**3.75**	**1.58**	**2.75**	**5.50**

Sd±Standard Deviation

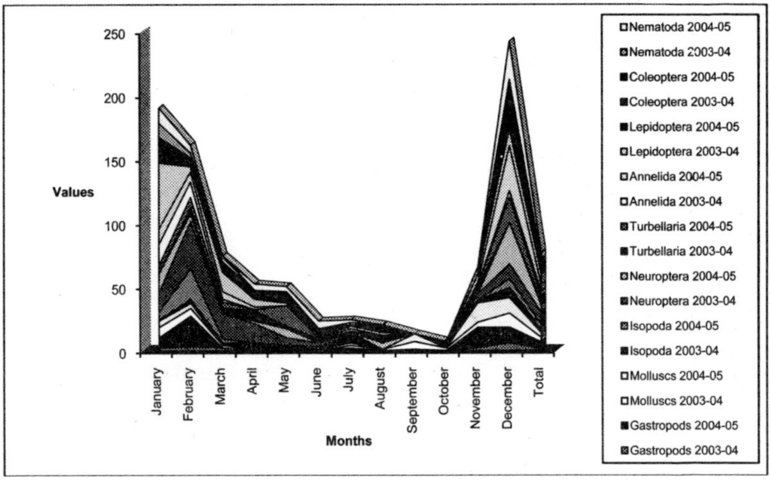

Fig. 4b.10: Graph showing values of different groups of macrobenthos at Sampling station B

Sampling station C

The comparative observation at sampling station A showed that the Turbellaria shown the dominancy over the other benthicfauna with the average value of 7.00 unit/sq.m. of the year 2003-2004. While Gastropods in 2004-2005, Molluscs in 2004-2005 and Annelida in 2004-2005 are observed minimum with average value of 1.42 unit/sq.m. (Table 4b.18 and figure 4b.23).

The Gastropods were observed to be minimum 0 unit/sq.m. in the month of June of the year 2003-2004 and was maximum 8 unit/sq.m. in the month of December of the year 2003-2004. The Molluscs were minimum 1 unit/sq.m. found in the months of October and November of the year 2003-2004 and the months June, September and November of the year 2004-2005 and was maximum 6 unit/sq.m. in the months of August of the year 2003-2004. The Isopoda were 1 unit/sq.m. in the months of February and September of the year 2003-2004 and was maximum 16 unit/sq.m. in December of the year 2003-2004. The Neuroptera found to be minimum 1 unit/sq.m. in the month of January of the year

2003-2004 and was maximum 8 unit/sq.m. in the month of February and March of the year 2004-2005. The Turbellaria were minimum 1 unit/sq.m. in the month of September of the year 2003-2004 and was maximum 32 unit/sq.m. in July and December of the year 2003-2004. The Annelida were minimum 0 unit/sq.m. in the months of February and March of the year 2003-2004 and was maximum 6 unit/sq.m. in January, July, November and December of the year 2003-2004. The Lepidoptera were minimum 2 unit/sq.m. in the February and March months of 2003-2004 and in September of 2004-2005 and was maximum 15 unit/sq.m. in January of the year 2004-2005. The total Coleoptera were observed to be minimum 1 unit/sq.m. in the month of February of the year 2003-2004, it was maximized in month of January with 32 unit/sq.m. of the year 2004-2005. The Nematoda were minimum 1 unit/sq.m. in the month of November of the year 2004-2005 and maximum 15 unit/sq.m. in the month of December of the year 2003-2004.

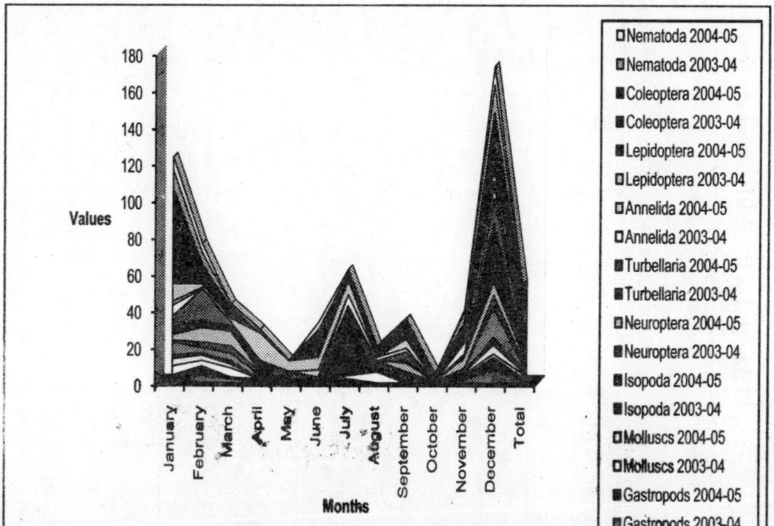

Fig. 4b.15: Graph showing values of different groups of macrobenthos at Sampling station C

Table 4b.18: Benthicfauna in the river Ganga during 2003-2005 at Sampling Station-C

Months	*Gastropods*		*Molluscs*		*Isopoda*		*Neuroptera*		*Turbellaria*		*Annelida*		*Lepidoptera*		*Coleoptera*		*Nematoda*	
	2003-04	2004-05	2003-04	2004-05	2003-04	2004-05	2003-04	2004-05	2003-04	2004-05	2003-04	2004-05	2003-04	2004-05	2003-04	2004-05	2003-04	2004-05
January	2	2	5	4	2	7	1	3	4	6	6	2	10	15	5	32	10	7
February	4	5	3	3	1	3	3	8	4	17	0	1	2	6	1	4	6	5
March	2	1	5	2	2	4	4	8	3	2	0	-	2	-	6	-	2	-
April	1	1	-	-	-	2	2	-	6	-	-	-	15	-	-	-	2	-
May	1	-	-	-	-	-	-	-	-	5	-	-	6	-	-	-	-	-
June	0	-	-	1	-	-	-	2	-	-	-	4	7	7	-	8	-	4
July	2	-	2	-	6	-	-	-	32	-	6	-	8	-	2	-	2	2
August	-	-	6	-	2	-	4	-	-	-	2	-	-	-	6	-	-	-
September	-	-	-	1	1	2	2	5	1	5	2	5	-	2	7	2	-	-
October	-	-	1	-	-	-	-	-	-	-	-	-	-	-	4	-	-	-
November	2	2	1	1	-	2	-	6	2	-	6	2	-	5	2	2	6	1
December	8	6	2	5	4	16	7	7	32	14	6	3	6	12	5	15	15	10
Total	**1.83**	**1.42**	**2.08**	**1.42**	**1.50**	**3.00**	**1.92**	**3.25**	**7.00**	**4.08**	**2.33**	**1.42**	**4.67**	**3.92**	**3.17**	**5.25**	**3.58**	**2.42**

Sd±Standard Deviation

Sampling station D

The comparative observation at sampling station A showed that the Gastropods shown the dominancy over the other benthicfauna with the average value of 6.50 unit/sq.m. of the year 2003-2004 and the Turebellaria with the minimum average value of 1.25 unit/sq.m. of the year 2003-2004 (Table 4b.19 and figure 4b.24).

The Gastropods were observed to be minimum 1 unit/sq.m. in the month of October of the year 2003-2004 and was maximum 32 unit/sq.m. in the month of January of the year 2003-2004.The Molluscs were minimum 1 unit/sq.m. found in the month of April of the year 2003-2004 and in the month May in the year 2004-2005 and was maximum 24 unit/sq.m. in the month of February of the year 2004-2005. The Isopoda were 1 unit/sq.m. in the months June, October, November and December of the year 2003-2004 and was maximum 12 unit/sq.m. in February of the year 2004-2005. The Neuroptera found to be minimum 1 unit/sq.m. in the months of February, March and October of the year 2003-2004 and was maximum 32 unit/sq.m. in the month of December of the year 2004-2005. The Turbellaria were minimum 1 unit/sq.m. in the months of February, May, October and November of the year 2003-2004 and in the month of April in the year 2004-2005. It was maximized to 35 unit/sq.m. in March month of the year 2004-2005. The Annelida were minimum 0 unit/sq.m. in the months of May, June and July of the year 2003-2004 and was maximum 32 unit/sq.m. in March of the year 2004-2005. The Lepidoptera were minimum 1 unit/sq.m. in the May and December month of 2003-2004 and was maximum 35 unit/sq.m. in December of the year 2004-2005. The total Coleoptera were observed to be minimum 0 unit/sq.m. in the months of February and May of the year 2003-2004 and it was maximized in month of January with 12 unit/sq.m. of the year 2004-2005. The Nematoda were minimum 0 unit/sq.m. in the months of February, May and November of the year 2003-2004 and maximum 12 unit/sq.m. in the month of April of the year 2003-2004.

Table 4b.19: Benthicfauna in the river Ganga during 2003-2005 at Sampling Station-D

Months	Gastropods		Molluscs		Isopoda		Neuroptera		Turbellaria		Annelida		Lepidoptera		Coleoptera		Nematoda	
	2003-04	2004-05	2003-04	2004-05	2003-04	2004-05	2003-04	2004-05	2003-04	2004-05	2003-04	2004-05	2003-04	2004-05	2003-04	2004-05	2003-04	2004-05
January	32	6	2	5	6	6	4	15	3	2	6	8	2	12	3	12	8	-
February	16	16	5	24	2	12	1	6	1	14	10	12	18	3	0	10	0	-
March	5	5	2	15	4	10	1	7	2	35	2	32	6	20	5	-	1	-
April	2	-	1	-	2	-	2	15	2	1	1	-	7	-	6	-	12	-
May	-	-	-	1	-	-	-	-	1	-	0	-	2	-	0	-	0	-
June	-	-	-	-	1	-	-	8	-	-	0	-	1	-	-	5	-	-
July	4	2	4	-	-	-	6	-	-	-	0	-	-	2	-	-	-	-
August	-	-	2	-	-	-	-	-	-	2	2	-	-	-	-	-	-	-
September	-	-	-	2	2	-	2	-	2	-	1	-	-	-	6	3	2	-
October	1	-	-	-	1	2	1	-	1	-	1	-	-	-	1	-	1	-
November	6	-	6	-	1	-	12	-	1	6	4	-	6	-	3	-	0	-
December	12	8	6	12	1	7	2	32	2	6	8	16	1	35	2	2	6	-
Total	**6.50**	**3.08**	**2.33**	**4.92**	**1.67**	**3.08**	**2.58**	**6.92**	**1.25**	**5.50**	**2.92**	**5.67**	**3.58**	**6.00**	**2.17**	**2.67**	**2.50**	**-**

Sd±Standard Deviation

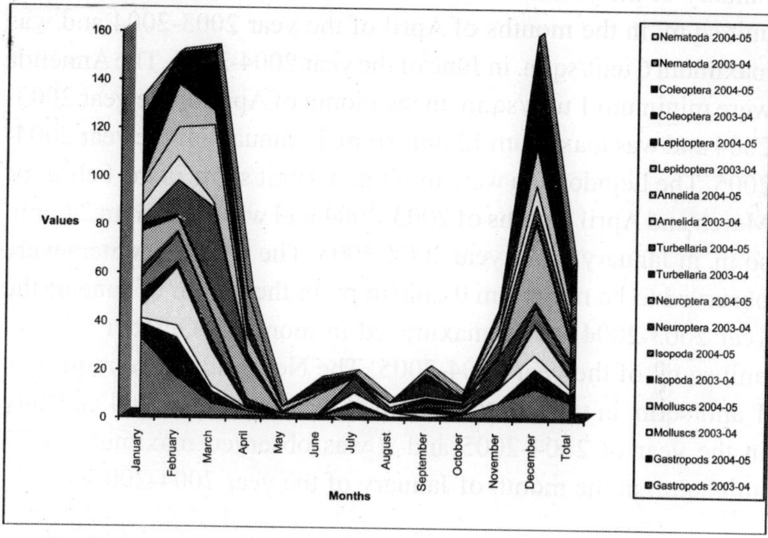

Fig. 4b.20: Graph showing values of different groups of macrobenthos at Sampling station D

Sampling station E

The comparative observation at sampling station A showed that the Neuroptera shown the dominancy over the other benthicfauna with the average value of 7.67 unit/sq.m. of the year 2004-2005 and the Turbellaria with the minimum average value of 0.92 unit/sq.m. of the year 2004-2005 (Table 4b.20 and figure 4b.25).

The Gastropods were observed to be minimum 1 unit/sq.m. in the month of February of the year 2003-2004 and in the month of May in the year 2004-2005. It was maximized 17 unit/sq.m. in the month of February of the year 2004-2005.The Molluscs were minimum 1 unit/sq.m. found in the month of March of the year 2003-2004 and was maximum 12 unit/sq.m. in the months of December of the year 2004-2005. The Isopoda were 1 unit/sq.m. in the month April of the year 2003-2004 and was maximum 32 unit/sq.m. in December of the year 2004-2005. The Neuroptera found to be minimum 1 unit/sq.m. in the month of May of the year 2003-2004 and was maximum 30 unit/sq.m. in the month of

January of the year 2004-2005. The Turbellaria were minimum 1 unit/sq.m. in the months of April of the year 2003-2004 and was maximum 6 unit/sq.m. in June of the year 2004-2005. The Annelida were minimum 1 unit/sq.m. in the month of April of the year 2003-2004 and was maximum 12 unit/sq.m. in January of the year 2004-2005. The Lepidoptera were minimum 0 unit/sq.m. in the February, March and April months of 2003-2004 and was maximum 24 unit/sq.m. in January of the year 2004-2005. The total Coleoptera were observed to be minimum 0 unit/sq.m. in the month of June of the year 2003-2004 it was maximized in month of October with 14 unit/sq.m. of the year 2004-2005. The Nematoda were minimum 1 unit/sq.m. in the month of June of the year 2003-2004 and July in the year of 2004-2005 and it was observed maximum of 37 unit/sq.m. in the month of January of the year 2004-2005.

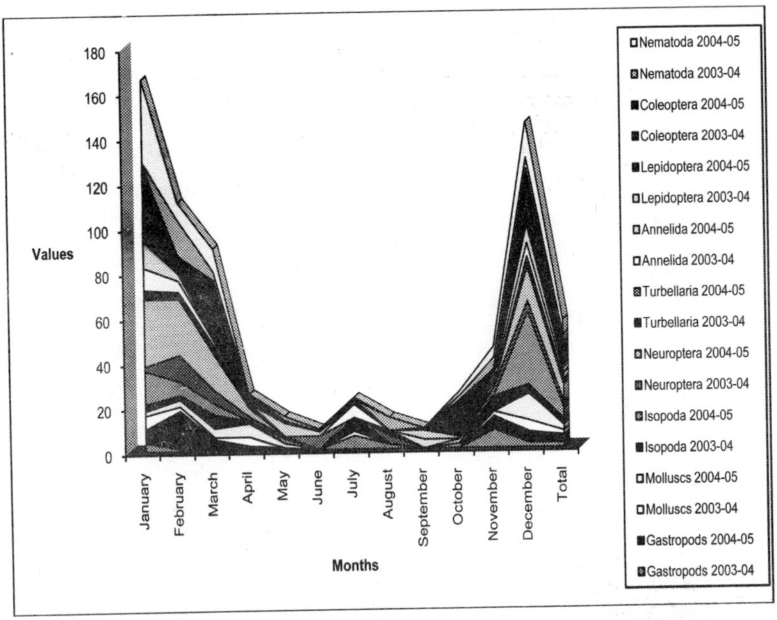

Fig. 4b.25: Graph showing values of different groups of macrobenthos at Sampling station E

Table 4b.20: Benthicfauna in the river Ganga during 2003-2005 at Sampling Station-E

Months	Gastropods		Molluscs		Isopoda		Neuroptera		Turbellaria		Annelida		Lepidoptera		Coleoptera		Nematoda	
	2003-04	2004-05	2003-04	2004-05	2003-04	2004-05	2003-04	2004-05	2003-04	2004-05	2003-04	2004-05	2003-04	2004-05	2003-04	2004-05	2003-04	2004-05
January	4	6	6	2	3	15	2	30	4	-	10	12	1	24	2	6	2	37
February	1	17	2	3	2	6	12	25	3	-	5	3	0	1	1	6	16	8
March	-	5	1	4	6	5	6	16	2	2	2	-	0	13	4	10	4	10
April	-	2	4	6	1	-	2	4	1	-	1	-	0	-	2	2	-	-
May	-	1	-	2	-	2	1	-	-	-	-	-	6	-	1	-	2	-
June	-	-	-	-	-	-	-	-	-	6	-	-	2	-	0	-	1	-
July	6	-	2	-	6	-	-	-	-	-	6	-	-	-	2	-	-	1
August	2	-	-	-	-	-	4	-	2	-	-	-	-	-	-	-	6	-
September	-	-	-	-	-	-	-	-	-	-	-	4	4	-	1	-	-	-
October	-	-	-	-	-	2	-	2	-	-	-	-	1	1	1	14	2	2
November	8	6	2	-	7	-	4	-	-	-	2	2	2	-	2	-	5	5
December	2	5	5	12	4	32	5	15	2	3	6	8	2	5	6	12	5	16
Total	**1.92**	**3.50**	**1.83**	**2.42**	**2.42**	**5.17**	**3.00**	**7.67**	**1.17**	**0.92**	**2.67**	**2.42**	**1.50**	**3.67**	**1.83**	**4.17**	**3.58**	**6.58**

Sd±Standard Deviation

23. Observation for icthyofauna (fish):

Fishes were collected from various spots of river Ganga all through the sampling range. The river contained rich fish diversity and population. As a result in the Ganga 44 species of fish were collected during the study period. Which are arranged systematically in Table 4b.21

Table 4b.21: Systematic list of fishes found in the river Ganga

Phylum	:	Vertebrata
Sub-phylum	:	Craniata
Super-class	:	Gnathostomata
Series	:	Piscean
Class	:	Telcostomi
Sub-class	:	Actinopterygiie
Order	:	Cypriniformes
Sub-order	:	Cyprinoidei
Family	:	Cyprinidae
Sub-family	:	Raborinae
Genus	:	Barilus Hamilton

Present scientific name	Name according to Day	Name according to Hamilton	Local name
Barilius bendelensis (Ham.)	Barilius bendelisis	*Cyprinus bendelisis*	Chal
Raimas bola (Ham.) (Endangered)	Barilius bola	*Cyprinus bola*	Bhola
Barilius vagra (Ham.)	Barilius vagra	*Cyprinus vagra*	Chal
-	Barilius modestus		
Barilius barna	-	-	
	Genus	*: Danio Hamilton*	
Danio (Danio) *devario* (Ham.)	Damino devario	*Cyprinius* devario	Patukari
	Sub-family	: Cyriniae	
	Genus	: Chaganio smith	
Chagunio chagunio (Ham)	Barbus chagunio	Cyprinius chagunio	Gelhari
	Genus	: Labeo Cuvier	
Labeo bogota (Ham.)	*L. bagota*	: *C. bogota*	Bata
Labeo gonius (Ham.)	*L. gonius*	: *C. gonius*	Rohu
Labeo rohita (Ham.)	*L. rohita*	: *C. rohita*	Kalbasu
Labeo calbasu (Ham.)	*L. calbasu*	*C. calbasu*	Aragi
Labeo dero (Ham.)	*L. dero*	*C. dero*	
Labeo dyocheilus (Mc. Clell)	*L. dyocheilus*	*L. dyocheilus*	
Labeo angra (Ham.)	*L. angra*	*L. angra*	Raiya
Eromus danricus (Ham.)	-	-	
Hypothalmichthys molitrix			
	Genus	: Crossocheilus vanhasselt	

(Contd...)

(...contd)

Crossocheilus latius latius (Ham.)	*Cirrhinus latia*	: *Cirrhinus latius*	Pet phorni
-	-	*C. gohan*	
-	-	*C. sada*	
-	*Cirrhinus mrigala*	*C. mrigala*	Nain
	Genus	: *Gara* Hamilton	
Garra gotyla gotyla (Gray)	*Discognathus lamata*	: *C. lamata*	Gunthale
	Genus	: *Tor gray*	
Tor putitora (Endangered)	*Barbus tor*	*C. tor*	Mahaseer
Tor tor (Ham.) (Endangered)	*Barbus tor*		*C. hexasticus*
Mahaseer			
Puntius hexasticus	Barbus hexasticus		
		Cyprinus sophore (Puntia)	Khront
Puntius sophore (Ham.)	Barbus sophore	-	Shidhari
Puntius sarna (Ham.)	*B. sarna*	*C. sarna*	Darahi
Puntius ticto (Ham.)	*B. ticto*	*C. ticto*	Sidhari
Puntius chola (Ham.)	*B. chola*	*C. chola*	Shidhari
Puntius conchonus (Ham.)			Darahi
	Sub-family	: Schyzothoracinae heckel	
	Genus	: Schizothorax heckel	
Schyzothorax plagiostomus	Oreinus plagiostomus		
Schyzothorax sinatus	Oreinus sinatus		

(Contd...)

(...contd)

	Family	: Cobitidae	
	Genus	: Botia gray	
Botia dario (Ham.)	*Botia dario*	Cabitus geto	Baghaua
Botia iohachata chandhuri	*B. iohachata chandhuri*	*Cabitus iohachata chandhuri*	
Lepdocephalus guntea (Ham.)	*L. guntea*	*L. guntea*	Hakati
	Genus	: Rita bleeker	
Rita rita (Ham.)	*Rita bachanai*	Pimelodus rita	Hunna rita
	Family	: Sisoridae	
	Genus	: Bagarius bleeker	
Bagarius bagarius	*Bagarius yarrelli*	Pimelodus sagarius	Gnoch
	Family	: Schilbeidae	
	Genus	: Clupisoma swainson	
Clupisoma garua (Ham.)	*Pseudeutropius garva*	Gilurus garua	Raikar karahi
	Order	: Beloiforms	
	Sub-order	: Scomerbosocoidei	
	Family	: Belonidae	
	Genus	: Xenetadon regan	
Xenetadon cancila (Ham.)	*Belon cancila*	Esor cancila	Kuwa
	Order	: Perciformes	
	Sub-order	: Anabantoidei	
	Family	: Anabanlidae	
	Genus	: Colisa cuver etvalenciennes	

(Contd...)

(...contd)

Closia fasciatuas (Block & Schn.)	*Trichogaster fasciatus*	*Trichopodius colisa*	Khosti
-	-	*T.bejeus*	
-	-	*T. corta*	
	Order	: *Mastacembeliformes*	
	Family	: *Mastacembelidas*	
	Genus	: *Mastacembelus schopoli*	
Mastacembelus armatus (Lacepede)	Mastacembelus armatus	*Macrognathus armatus*	Bam
Mastacembelus punctatus (Ham.)	Mastacembelus punctatus	*Macrognathus punctatus*	Malga
	Family	: Bragridae	
Mystus seenghala (Sykes)	-	-	Tengar
Mystus bleekeri (Day)	-	-	Tengar
Mystus cavasius (Ham.)	-	-	Tengar
Mystus tengara (Ham.)	-	-	Tengar
Mystus vittatus (Bl.)	-	-	Tengar
Mystus aor (Ham.)	-	-	Tengar
	Family	: Nadidae	
Nandus nandus (Ham.)	Nandus marmoratus	Gobius nandus	Dhebari

(4.c) MICROBIAL EXAMINATION:

Sampling station A: -

The MPN (/100ml) of the Ganga river at this sampling spot was recorded minimum 700.00 MPN/ 100 ml ± 15.00 in the month of January in the year 2004-2005 and maximum of 8600.00 MPN/ 100 ml ± 72.00 in the month of July in the year 2004-2005. The average value in the year 2003-2004 was 2698.83 MPN/ 100 ml ± 45.33 and 2658.33 MPN/ 100 ml ± 29.83 in the year 2004-2005. (Table 4c.1 and Figure 4c.1- 4c.2)

The SPC/ml (×103) of Sampling station A was recorded minimum 10.00 × 103 SPC/ ml ± 1.00 in February month of the year 2004-2005 and maximum of 200.00 × 103 SPC/ ml ± 3.00 in the month of July in the year 2004-2005. The average value in the year 2003-2004 was 63.77 × 103 SPC/ ml ± 2.01 and was recorded to be 97.42 × 103 SPC/ ml ± 2.45 in the year2004-2005. (Table 4c.1 and Figure 4c.3 - 4c.4)

Table 4c.1: Results of microbiologial characteristic at sampling station A during 2003-2005

Months	Total Coliform- MPN/ 100 ml. Microbial Parameter		SPC/ml ($x10^3$)	
	2003-2004	2004-2005	2003-2004	2004-2005
January	1126.00±76.00	700.00±15.00	15.00±0.40	16.00±2.00
February	876.00±15.00	1200.00±35.00	12:00±0.30	10.00±1.00
March	968.00±24.00	1100.00±18.00	17.00±1.80	25.00±3.00
April	1560.00±40.00	2500.00±30.00	32.00±2.00	72.00±2.00
May	2600.00±32.00	2000.00±20.00	56.00±5.00	80.00±3.00
June	2500.00±25.00	2500.00±40.00	116.00±2.80	110.00±6.20
July	1800.00±30.00	8600.00±72.00	172.00±2.40	200.00±3.00
August	7800.00±78.00	6500.00±52.00	106.20±3.40	160.00±2.60
September	6500.00±65.00	3200.00±15.00	98.00±2.00	140.00±2.00
October	3200.00±56.00	1600.00±10.00	76.00±1.00	136.00±2.00
November	2400.00±45.00	1200.00±16.00	45.00±2.10	200.00±1.00
December	1056.00±58.00	800.0±35.00	20.00±0.90	20.00±1.60
Average	**2698.83±45.33**	**2658.33±29.83**	**63.77±2.01**	**97.42±2.45**

Sd± Standard Deviation

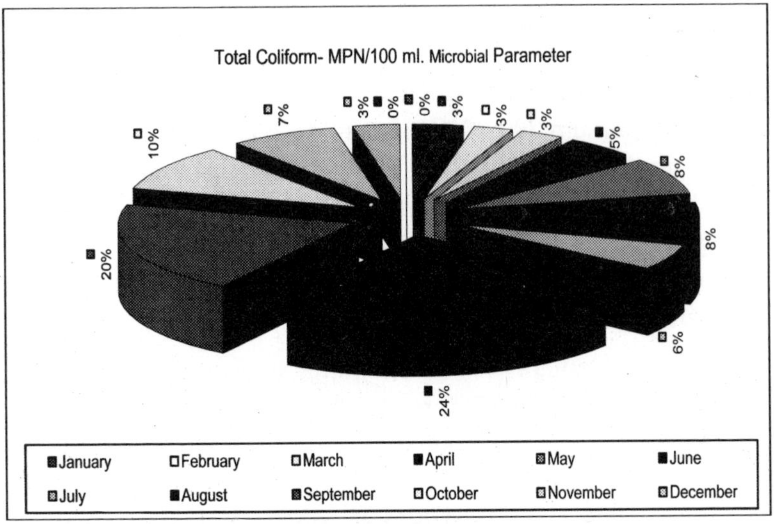

Fig. 4c.1: Graph showing percentage of total coliform for 2003-2004 at Sampling station A

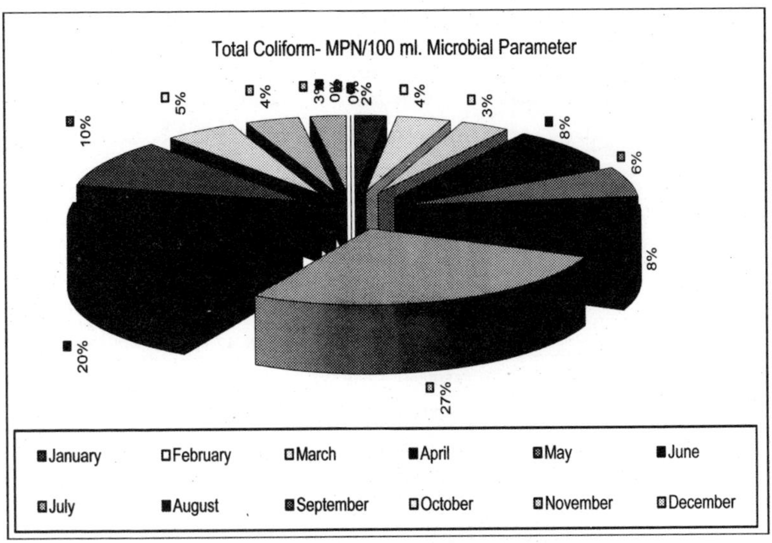

Fig. 4c.2: Graph showing percentage of total coli form for 2004-2005 at Sampling station A

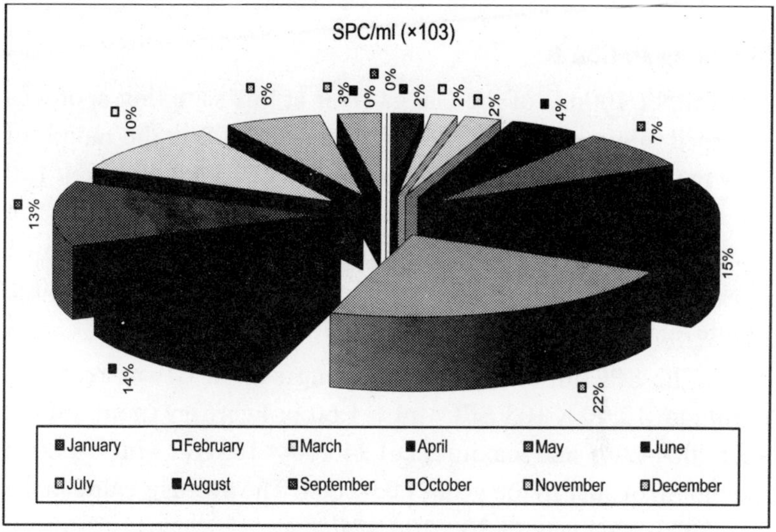

Fig. 4c.3: Graph showing percentage of SPC for 2003-2004 at Sampling station A

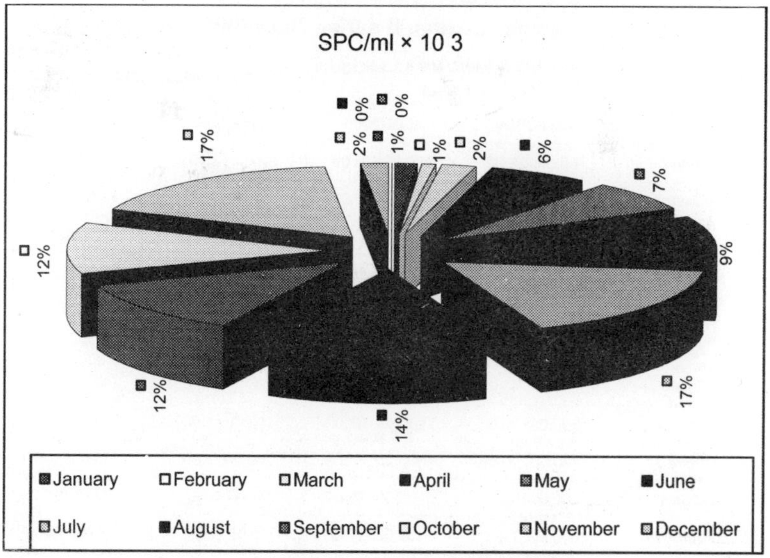

Fig. 4c.4: Graph showing percentage of SPC for 2004-2005 at Sampling station A

Sampling station B: -

The MPN (/100ml) of the Ganga river at this sampling spot was recorded minimum 27.00 MPN/ 100 ml ± 25.00 in the month of May in the year 2004-2005 and maximum of 15000.00 MPN/ 100 ml ± 217.00 in the month of July in the year 2004-2005. The average value in the year 2003-2004 was 2143.33 MPN/ 100 ml ± 17.43 and 3497.50 MPN/ 100 ml ± 46.92 in the year 2004-2005 (Table 4c.2 and Figure 4c.5-4c.6).

The SPC/ml (×103) of Sampling station B was recorded minimum 12.00 × 103 SPC/ ml ± 1.60 in February month of the year 2004-2005 and maximum of 240.00 × 103 SPC/ ml ± 4.80 in the month of July in the year 2003-2004. The average value in the year 2003-2004 was 83.50 × 103 SPC/ ml ± 7.50 and was recorded to be 71.17 × 103 SPC/ ml ± 1.86 in the year2004-2005 (Table 4c.2 and Figure 4c.7-4c.8).

Table4c.2: Results of microbiologial characteristic at sampling station B during 2003-2005

Months	Total Coliform- MPN/100 ml. Microbial Parameter		SPC/ml ($x10^3$)	
	2003-2004	2004-2005	2003-2004	2004-2005
January	760.00±12.00	900.00±16.00	35.00±2.40	16.00±0.20
February	540.00±12.00	800.00±20.00	16.00±2.10	12.00±1.60
March	9870.00±20.00	1250.00±35.00	45.00±3.60	65.00±2.50
April	1540.00±20.00	1500.00±36.00	95.00±35.00	70.00±3.00
May	1600.00±12.00	27.00±25.00	84.00±21.00	50.00±2.60
June	2500.00±30.00	3200.00±60.00	146.00±2.70	32.00±1.60
July	3760.00±40.00	15000.00±217.00	240.00±4.80	116.00±2.60
August	1600.00±30.00	7500.00±62.00	106.00±2.00	170.00±2.00
September	1200.00±10.00	4500.00±24.00	102.00±3.00	140.00±1.60
October	900.00±5.10	2400.00±26.00	98.00±2.40	127.00±1.20
November	600.00±2.00	1200.00±17.00	15.00±1.00	26.00±1.60
December	850.00±16.00	1020.00±25.00	20.00±10.00	30.00±1.80
Average	**2143.33±17.43**	**3497.50±46.92**	**83.50±7.50**	**71.17±1.86**

Sd± Standard Deviation

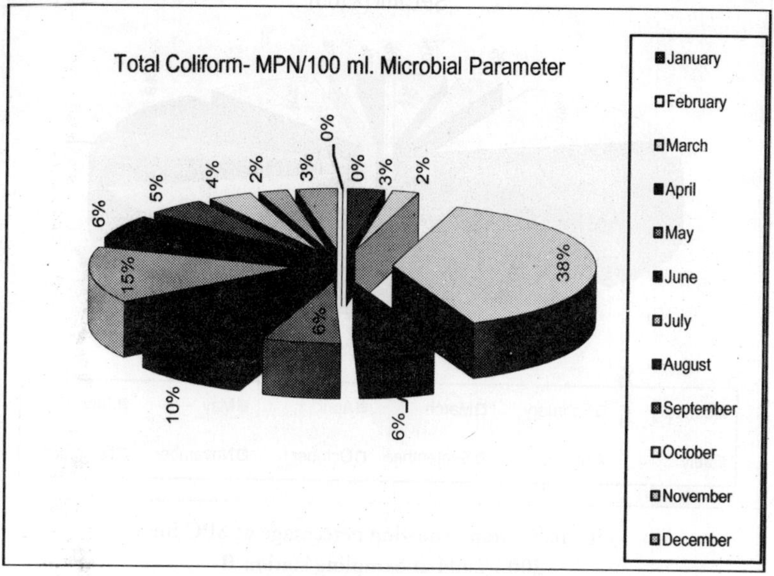

Fig. 4c.5: Graph showing percentage of total coliform for 2003-2004 at Sampling station B

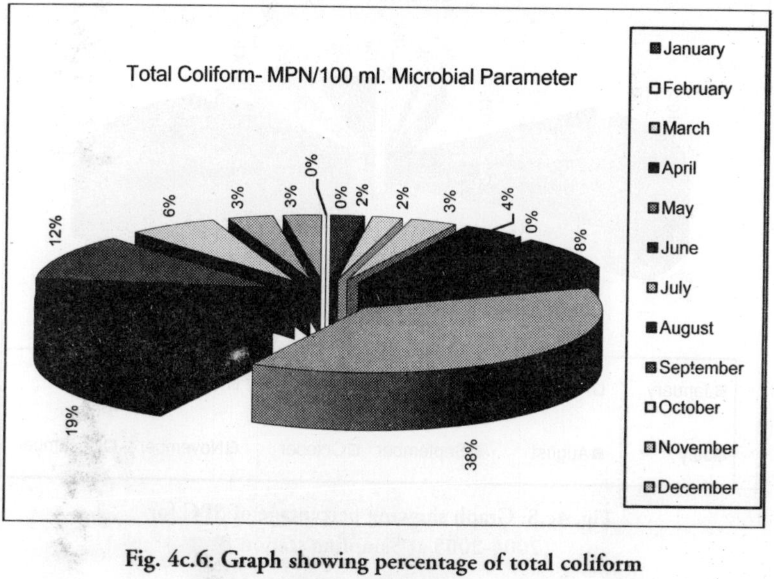

Fig. 4c.6: Graph showing percentage of total coliform for 2004-2005 at Sampling station B

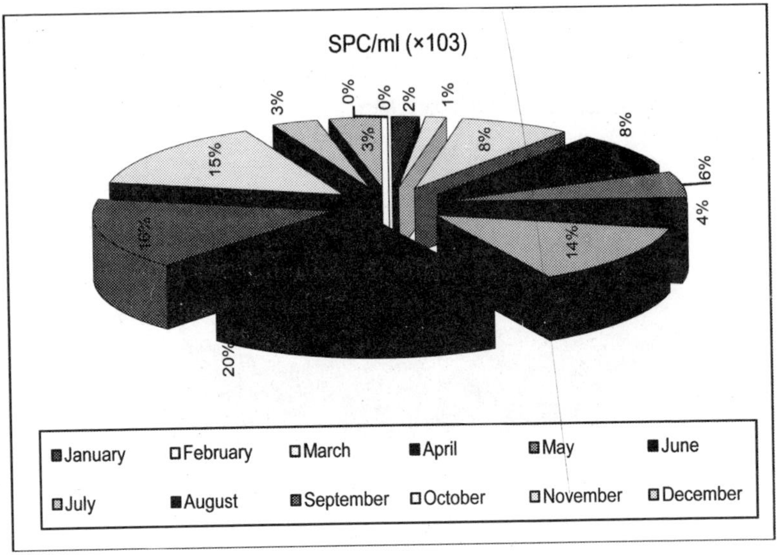

Fig. 4c.7: Graph showing percentage of SPC for 2003-2004 at Sampling station B

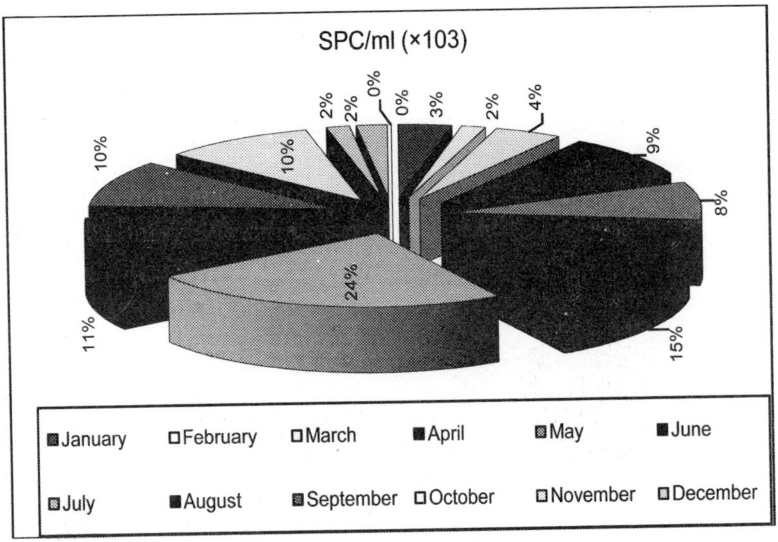

Fig. 4c.8: Graph showing percentage of SPC for 2004-2005 at Sampling station B

Sampling station C: -

The MPN (/100ml) of the Ganga river at this sampling spot was recorded minimum 800.00 MPN/ 100 ml ± 27.00 in the month of January in the year 2003-2004 and maximum of 25000.00 MPN/ 100 ml ± 70.00 in the month of July in the year 2004-2005. The average value in the year 2003-2004 was 2063.33 MPN/ 100 ml ± 28.33 and 4651.67 MPN/ 100 ml ± 35.42 in the year 2004-2005 (Table 4c.3 and Figure 4c.9 - 4c.10).

The SPC/ml (×103) of Sampling station C was recorded minimum 7.00 × 103 SPC/ ml ± 1.00 in February month of the year 2003-2004 and maximum of 262.00 × 103 SPC/ ml ± 2.50 in the month of July in the year 2004-2005. The average value in the year 2003-2004 was 69.00 × 103 SPC/ ml ± 2.03 and was recorded to be 108.29 × 103 SPC/ ml ± 3.06 in the year 2004-2005. (Table 4c.3 and Figure 4c.11 - 4c.12)

Table 4c.3: Results of microbiologial characteristic at sampling station C during 2003-2005

Months	Total Coliform- MPN/100 ml. Microbial Parameter		SPC/ml (x10^3)	
	2003-2004	2004-2005	2003-2004	2004-2005
January	800.00±27.00	900.00±25.00	16.00±0.50	25.00±1.60
February	1200.00±16.00	1600.00±21.00	7.00±1.00	37.00±2.00
March	1700.00±18.00	1800.00±30.00	25.00±2.00	40.50±3.50
April	2200.00±40.00	1600.00±20.00	72.00±3.00	82.00±2.60
May	2400.00±46.00	2400.00±25.00	86.00±2.80	90.00±7.00
June	3200.00±40.00	5200.00±60.00	116.00±5.70	172.00±8.00
July	5600.00±58.00	25000.00±70.00	172.00±3.00	262.00±2.50
August	3100.00±32.00	8400.00±60.00	116.00±2.40	216.00±2.00
September	1500.00±12.00	3200.00±24.00	102.00±2.00	160.00±3.60
October	1200.00±10.00	2400.00±40.00	76.00±1.00	120.00±2.00
November	1000.00±6.00	1600.00±20.00	20.00±0.20	75.00±1.70
December	860.00±35.00	1720.00±30.00	20.00±0.80	20.00±0.20
Average	2063.33±28.33	**4651.67±35.42**	**69.00±2.03**	**108.29±3.06**

Sd± Standard Deviation

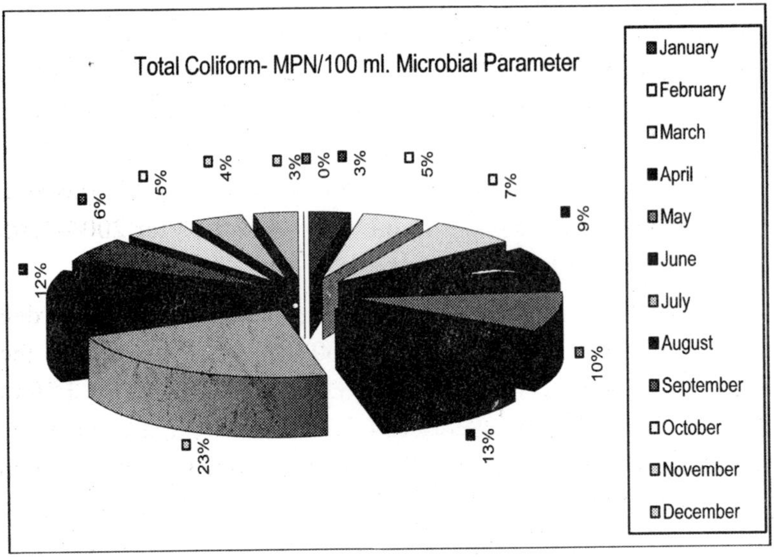

Fig. 4c.9: Graph showing percentage of total coliform for 2003-2004 at Sampling station C

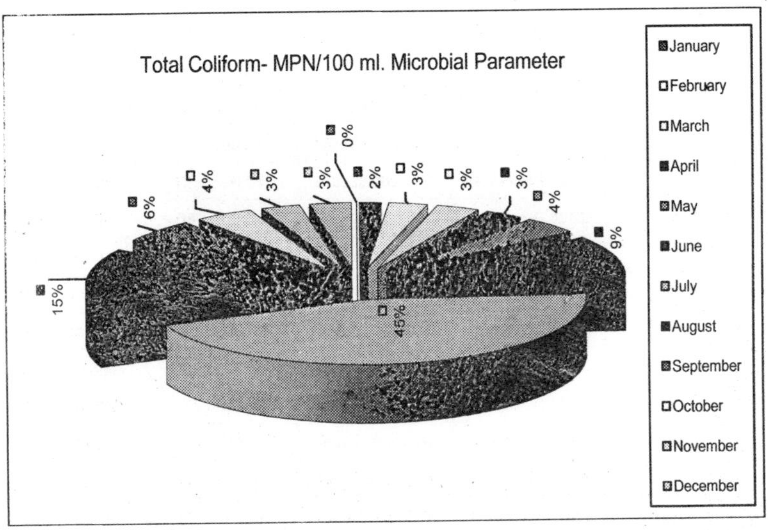

Fig. 4c.10: Graph showing percentage of total coliform for 2004-2005 at Sampling station C

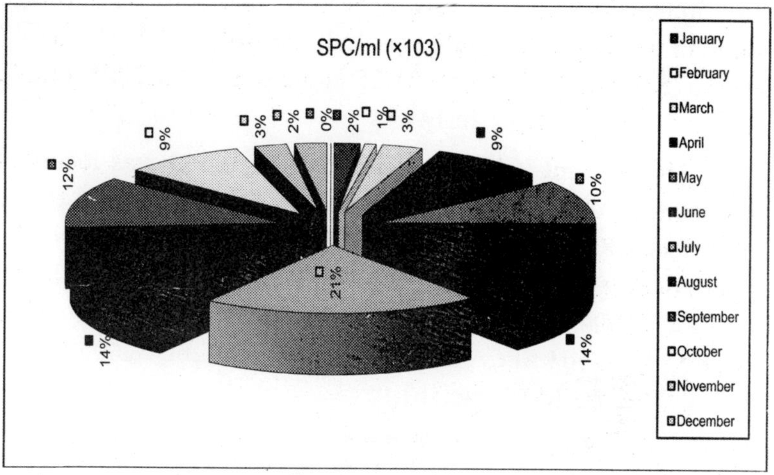

Fig. 4c.11: Graph showing percentage of SPC for 2003-2004 at Sampling station C

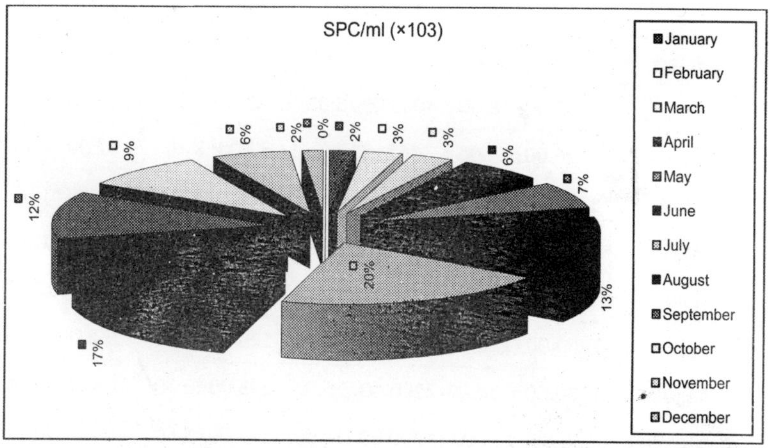

Fig. 4c.12: Graph showing percentage of SPC for 2004-2005 at Sampling station C

Sampling station D: -

The MPN (/100ml) of the Ganga river at this sampling spot was recorded minimum 25.00 MPN/ 100 ml ± 30.00 in the month of May in the year 2003-2004 and maximum of 4500.00 MPN/ 100 ml ± 30.00 in the month of July in the year 2004-2005. The average value in the year 2003-2004 was 1543.08 MPN/ 100 ml ± 50.42 and 1675.58 MPN/ 100 ml ± 21.25 in the year 2004-2005. (Table 4c.4 and Figure 4c.13 - 4c.14)

The SPC/ml (×103) of Sampling station D was recorded minimum 16.00 × 103 SPC/ ml ± 0.20 in February month of the year 2004-2005 and maximum of 235 × 103 SPC/ ml ± 6.10 in the month of August in the year 2003-2004. The average value in the year 2003-2004 was 88.42 × 103 SPC/ ml ± 3.33 and was recorded to be 83.92 × 103 SPC/ ml ± 2.20 in the year 2004-2005 (Table 4c.4 and Figure 4c.15 - 4c.16).

Table 4c.4: Results of microbiologial characteristic at sampling station D during 2003-2005

Months	Total Coliform- MPN/100 ml. Microbial Parameter		SPC/ml ($\times10^3$)	
	2003-2004	2004-2005	2003-2004	2004-2005
January	1260.00±12.00	670.00±10.00	50.00±2.60	17.00±1.60
February	1200.00±18.00	800.00±25.00	72.00±1.90	16.00±0.20
March	1526.00±25.00	862±32.00	32.00±2.10	32.00±2.40
April	1600.00±15.00	1600.00±25.00	60.00±3.16	52.00±1.60
May	25.00±30.00	2500.00±30.00	58.00±5.00	50.00±1.00
June	1200.00±20.00	2700.00±32.00	68.00±2.00	162.00±3.60
July	2800.00±125.00	4500.00±30.00	126.00±5.60	168.00±3.80
August	1700.00±120.00	2100.00±25.00	235.00±6.10	150.00±2.60
September	856.00±125.00	1700.00±12.00	125.00±3.60	140.00±5.00
October	2500.00±50.00	1200.00±10.00	110.00±2.00	125.00±2.40
November	2100.00±30.00	850.00±12.00	75.00±2.10	75.00±1.60
December	1750.00±35.00	625.00±12.00	50.00±3.80	20.00±0.60
Average	**1543.08±50.42**	**1675.58±21.25**	**88.42±3.33**	**83.92±2.20**

Sd± Standard Deviation

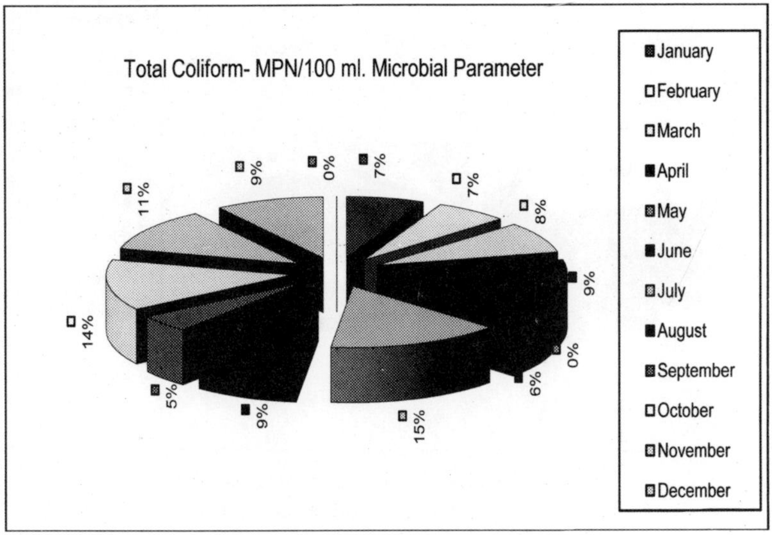

Fig. 4c.13: Graph showing percentage of total coliform for 2003-2004 at Sampling station D

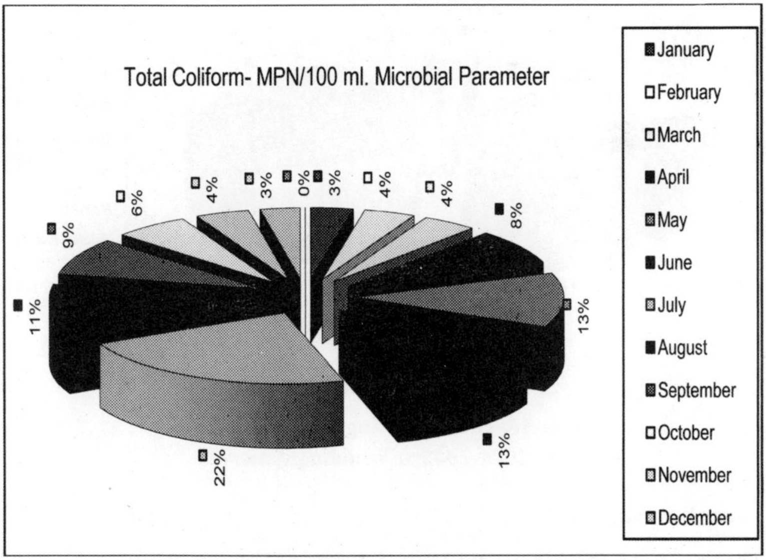

Fig. 4c.14: Graph showing percentage of total coliform for 2004-2005 at Sampling station D

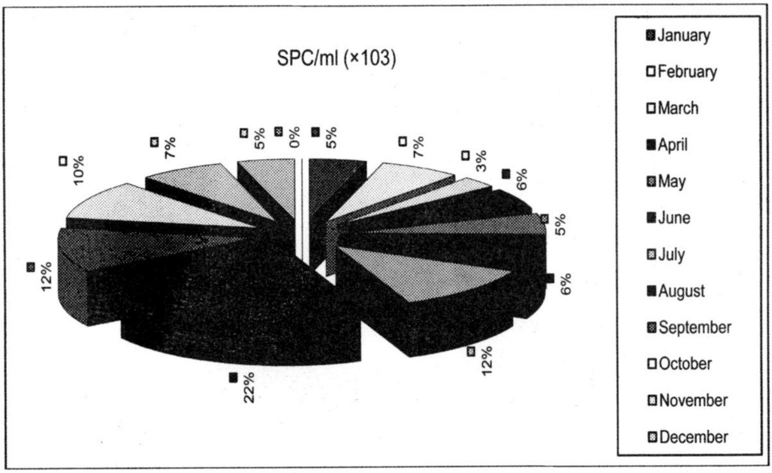

Fig. 4c.15: Graph showing percentage of SPC for 2003-2004 at Sampling station D

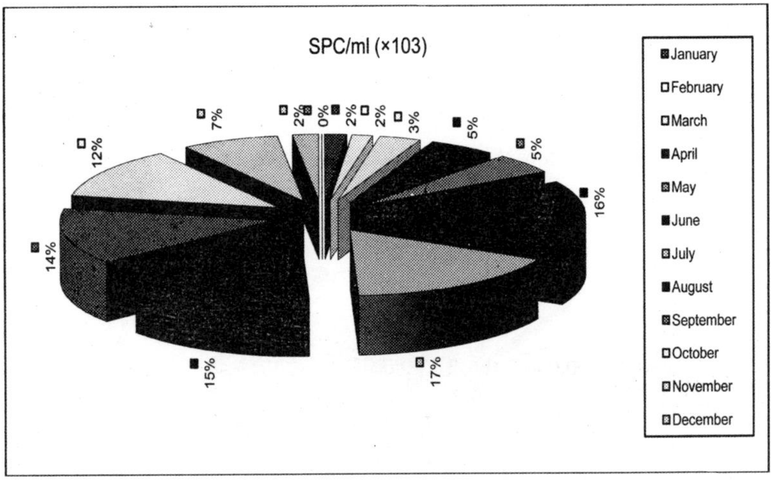

Fig. 4c.16: Graph showing percentage of SPC for 2004-2005 at Sampling station D

Sampling station E :-

The MPN (/100ml) of the Ganga river at this sampling spot was recorded minimum 765.00 MPN/ 100 ml ± 25.00 in the month of

February in the year 2003-2004 and maximum of 27000.00 MPN/ 100 ml ± 80.00 in the month of July in the year 2004-2005. The average value in the year 2003-2004 was 3752.08 MPN/ 100 ml ± 67.75 and 4908.33 MPN/ 100 ml ± 36.00 in the year 2004-2005. (Table 4c.5 and Figure 4c.17 - 4c.18)

The SPC/ml (x103) of Sampling station E was recorded minimum 16.00 × 103 SPC/ ml ± 1.20 in February month of the year 2004-2005 and maximum of 260.00 × 103 SPC/ ml ± 5.00 in the month of July in the year 2004-2005. The average value in the year 2003-2004 was 61.63 × 103 SPC/ ml ± 1.66 and was recorded to be 104.33 × 103 SPC/ ml ± 2.07 in the year 2004-2005. (Table 4c.5 and Figure 4c.19 - 4c.20)

Table 4e.5: Results of microbiologial characteristic at sampling station E during 2003-2005

Months	Total Coliform- MPN/100 ml. Microbial Parameter		SPC/ml ($\times 10^3$)	
	2003-2004	2004-2005	2003-2004	2004-2005
January	850.00±35.00	900.00±12.00	46.50±1.68	25.00±1.60
February	765.00±25.00	1400.00±32.00	35.00±1.00	16.00±1.20
March	1650.00±70.00	3200.00±52.00	17.00±0.60	24.00±2.00
April	1500.00±46.00	3600.00±50.00	25.00±1.50	78.00±3.10
May	2100.00±75.00	2400.00±20.00	45.00±1.30	82.00±2.60
June	1600.00±60.00	4000.00±50.00	60.00±2.80	172.00±3.00
July	19000.00±215.00	27000.00±80.00	65.00±2.80	260.00±5.00
August	7500.00±62.00	5500.00±60.00	196.00±3.20	225.00±2.40
September	4100.00±50.00	4500.00±25.00	140.00±2.20	165.00±1.60
October	2700.00±68.00	2700.00±16.00	78.00±1.60	130.00±1.20
November	1500.00±40.00	1600.00±10.00	32.00±1.20	56.00±0.50
December	1760.00±67.00	2100.00±25.00	32.15±0.18	19.00±0.60
Average	**3752.08±67.75**	**4908.33±36.00**	**61.63±1.66**	**104.33±2.07**

Sd± Standard Deviation

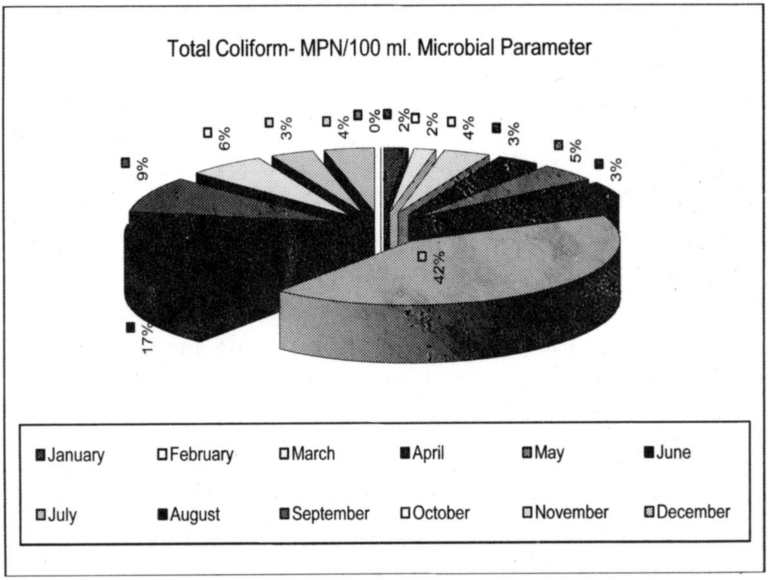

Fig. 4c.17: Graph showing percentage of total coliform for 2003-2004 at Sampling station E

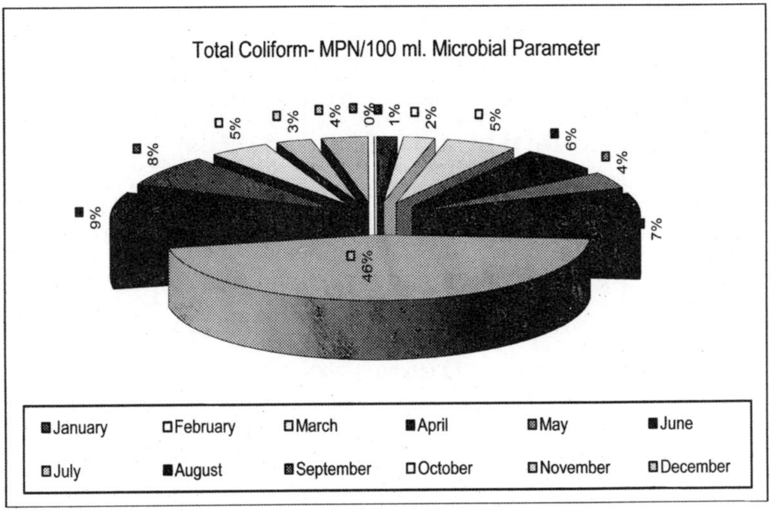

Fig. 4c.18: Graph showing percentage of total coliform for 2004-2005 at Sampling station E

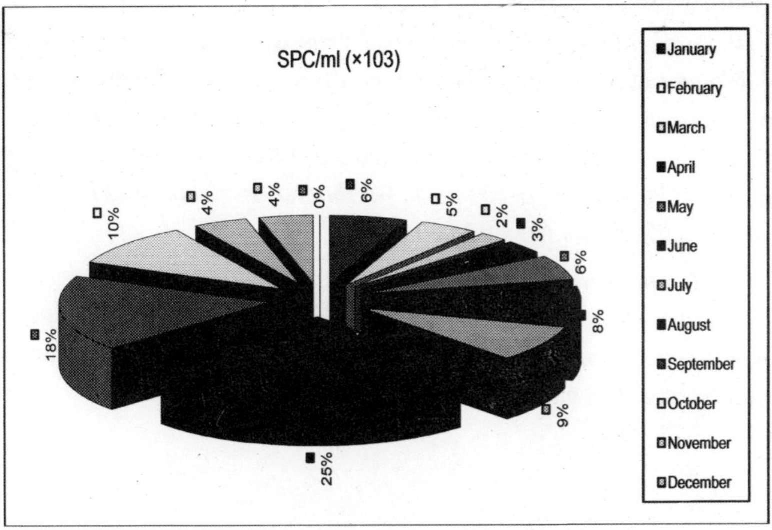

Fig. 4c.19: Graph showing percentage of SPC for 2003-2004 at Sampling station E

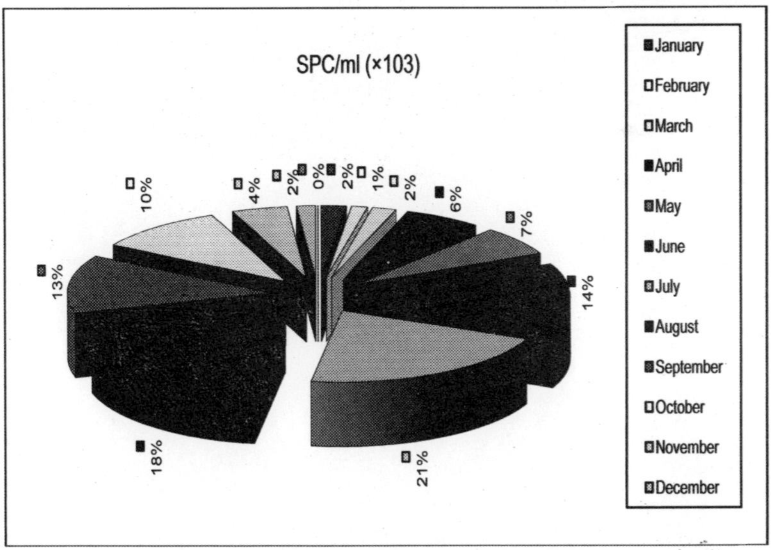

Fig. 4c.20: Graph showing percentage of SPC for 2004-2005 at Sampling station E

Identification of Bacteria :

In the present study 7 species of bacteria were identified in which pathogens and non-pathogens were differentiated within the stretch of study area. Isolated bacterial cultures were *Escherichia coli, Pseudomonas aeruginosa, Streptococcus facecalis, Staphylococcus aureus, Clostridium perfringes, Aerobacter aerogens,* and *Bacillus subtilis*. (Table 4c.6).

Escherichia coli, Pseudomonas aeruginosa, Streptococcus facecalis, Staphylococcus aureus, Clostridium perfringes, Aerobacter aerogens were found as pathogens whereas the non-pathogenic bacteria were *Bacillus subtilis*. (Table 4c.6).

CHAPTER-5

DISCUSSION

The present study is devoted for evaluation of different physical, chemical, biological and microbiolgical parameters of the river Ganga.

The temperature is one of the most important parameter in an aquatic environment. In fact, no other single factor has so intense influence and direct and indirect effect on the biota of an ecosystem. In the present study, a difference in the fluctuation of water temperature was observed 10.00°C ± 0.00 (minimum) in the month of November and 25.75°C ± 1.07 (maximum) in March. Annual average values of temperature varied between 14.55°C ± 1.78 to 17.55°C ± 0.44. Minimum annual average value of temperature observed in the year 2004-2005 at sampling station A and maximum was observed in the year 2003-2004 at Sampling Station A. The water temperature showed an upward trend from January to April followed by a downward from May onwards. A more or less similar trend has been observed in the river Yamuna by Chakrabarty *et al.* (1959) and in the Kallayi (John, 1976). Badola and Singh (1981) also reported similar trend in the river Alaknanda. Same study was made by Singh *et al.* (1988; 1989) in river Ganga, Yamuna and Sangam and Gautam *et al.* (2000) in Ganga river at Rishikesh. Same trend of temperature was observed by Khanna *et al.* (2001) in river Ganga at Haridwar.

The electrical conductivity has always been used as a valuable method to estimate the degree of salt contents and total dissolved solids in water. The present study showed conductivity fluctuation of 32.11 Simens/cm ± 0.24 to 360.07 Simens/cm ± 0.00. It can be said that the present higher conductivity values in August to October month due to the input of large amount of salts and silt carried by the river. Identical results were observed by Das *et al.*, (1992), Bhattacharya and Saha (1988), Raina *et al.*, (1982) and Mittal and Sengar (1990) from various Indian rivers. Khanna *et al.* (2003) also observed similar trend of conductivity in Ganga river at Bulandshahar.

The water of the river Ganga becomes start turbid from June month onward and in July to September the water was highly turbid. It showed that turbidity and total solids were closely inter related with one another and cause common effect upon the river and aquatic life as also stated by Verma *et al.* (1984). In the month of February and March some sampling station recorded Nil turbidity and highest (1264.00 JTU ± 13.752) noted in the month of September. Annual average value of turbidity varied between 67.59 JTU ± 8.61 to 177.90 JTU ± 21.39 in which maximum average value observed at sampling station D in the year 2003-2004 and minimum at sampling station B in 2004-2005. Similar pattern was also reported by Badola and Singh (1981), Dobriyal *et al.* (1983) in the hill streams of the Garhwal Himalaya, Khanna *et al.* (1997) in river Ganga, and Seth *et al.* (2000) in the river Ganga. The turbidity is a striking characteristic to know the physical status of the rivers. The suspended particles, soil particles, discharged effluents; decomposed organic matters, total dissolved solids as well as the microscopic organisms increase the turbidity of water, which interferes with the penetration of light. The increased turbidity during monsoon season was attributed due to heavy soil erosion in the nearby catchments and massive contribution of suspended solids from sewage canals. Surface runoffs and domestic wastes mainly contribute to the increased turbidity of the river. Bhatt *et al.* (1984) observed low turbidity during November, February and an increase after May and reaches peak in August in river Kosi. He attributed that during monsoon months, the water contained large amount of silt, fine sand particles, organic matter and clay. In river Ganga and its tributaries, minimum values for turbidity were recorded in winter and summer season while maximum in monsoon period (Bilgrami and Duttamunshi, 1985). Higher values were recorded at places where human activity was intense or at places where effluents from industries/Farms are discharged into the river. High amount of silt carried due to fast currents of Kosi attributed mainly for the turbidity of this river (Bhatt *et al.*, 1984).

The Maximum velocity (3.90m/sec. ± 0.75) recorded in July month and minimum of 0.10 m/sec.± 0.00 in the month of March. The annual average was ranges between 0.84 m/sec.± 0.13 to 1.87 m/sec.± 0.52. The velocity started increasing from May onwards due to melting of snow at the place of origin of the river Ganga. Again beyond September, the velocity started decreasing. In the present study the velocity and turbidity showed a direct relationship. Similar observations have been recorded by Dobriyal (1985) and Nautiyal (1990) in the snow fed rivers of the Garhwal Himalaya.

In the present investigation it was noted that the total solids were maximum in the month of June (1793.68 mg/l ± 0.16), which were responsible for the turbidity in the river. The total solids were recorded minimum (1.50 mg/l ± 0.82) in the month of February due to gradual sedimentation of the filterable residue. Annual average values of total solids varied between 334.61 mg/l ± 3.00 to 475.94 mg/l ± 5.03. Minimum average value of total solids observed in the year 2003-2004 at sampling station E and maximum in the year 2003-2004 at sampling station A. T.D.S. (Total dissolved solid) a part of total solid is also behave like total solid. In our study T.D.S. found minimum of 62.15 mg/l ± 0.20 in the month of March and the maximum upto 895.65 mg/l ± 2.54 in the month of July, with the annual average ranges between 122.80 mg/l ± 12.29 to 270.82 mg/l ± 3.97. The second contributor for total solid is Total suspended solid having the maximum and minimum value range observed during out study was 1476.08 mg/l ± 0.10 in the month of June to 1.90 mg/l ± 60.86 in November. Khanna and Singh (2000) found that total solids were maximum in summer in the water of Suswa river. Total solids cause ecological imbalance in the aquatic ecosystem by mechanical abrasive action. Higher values of total solids may cause deterioration of the surviving condition of aquatic organisms by mechanical abrasive action and enhance the turbidity of the river. Same conditions were shown by Khanna (1993), Verma & Shukla (1969) in their studies. Zingde *et al.* (1980), Kudesia and Verma (1985) and Reddy and Venkateswarlu (1987) studied that most of the Indian rivers show similar tendency with respect to fluctuations of total solids. Khanna *et al.* (1997) and Seth *et al.* (2000) also made out the same study. Similar trends were shown by Khanna and Chugh (2004) during study of water quality of River Ganga at Haridwar. Minimum values of total dissolved solids were obtained during winter season (34.00 mg/l ± 1.67) and maximum values during monsoon season (1200.00 mg/l ± 142.75). Annual average value of total dissolved solids varied between 105.24 mg/l ± 13.10 to 787.93 mg/l ± 326.17, in which minimum average value obtained from sampling station C in 2002-2003 and maximum from sampling station D in the year 2003-2004.

Absolutely pure water is weakly dissociated and neutral in reaction i.e. H^+ or OH^- ions will be present in equal proportions. So since the dissociation constant remains the same, the concentration of opposite ions changes simultaneously. Limnologists in studies of fresh water systems have assiduously measured the pH. Hutchinson (1941) doubted that there had been a serious demonstration that pH in itself influences the natural distribution and abundance of aquatic animals, with the exception of few protozoan. It has somehow an indirect relation with natural life prevailing in water system. Fishes are relatively said to be poor regulators of internal pH, thus change in pH of their environment can alter enzyme activities or electrolyte composition of their body fluid, providing severe stress. Low pH also interferes with oxygen uptake, and pH outside range of 4.00 to 10.00 can kill fish (Mathews, 1998). The water of Ganga river at Haridwar was found to be slightly alkaline during the course of study. The pH was maximum (8.90 ± 0.07) in November at sampling station E may be due to increased ashes from nearby ashrams and washing activities. Higher concentration of calcium in water may also increase pH. Lal (1996) also found the effects of mass bathing on water quality of Pushkar Sarovar. The pH was recorded to be minimum (5.94) in the month of February at sampling station E. Similar trend have been reported by Badola (1979). However, in contrast to this, Khanna and Chugh (2004) observed higher pH value in winter season at Bhimgoda. Khanna (1993) also reported higher pH values in winter and less in Rainy season. In almost all sampling stations the pH was slightly alkaline.

Maximum dissolved oxygen was recorded (14.35 mg/l ± 0.00) in the month of December at the sampling station E. The minimum value of dissolved oxygen (6.17 mg/l ± 0.12) was observed in the month of June at sampling station E. The annual average value of dissolved oxygen ranged between 9.73 mg/l ± 0.13 to 8.90 mg/l ± 0.73 whereas the minimum annual average value of DO was observed in the year 2004-2005 at sampling station A and maximum was observed in 2003-2004 at sampling station B. The dissolved oxygen reduced gradually from May onward due to turbidity which retarded the photosynthetic activity of aquatic flora. The temperature showed an inverse relationship with the DO almost throughout the study. The cause of maximum dissolved oxygen in November to February is due to reduced rate of decomposition by decreased microbial activity at low temperature (Strommer and Smock, 1989). Chopra *et al.* (1990), Gopal and Sah (1993) and Sharma (1999) also have got the same result and have opined that low temperature in November to February increases the oxygen retaining capacity of water and solubility of O_2 in water. This trend was also observed by Badola and Singh (1981) in the river Alaknanda. Khanna (1993) has also reported the same trends in the river Ganga at Haridwar. Same study is also made by Pandey *et al.* (2003) in Ganga canal at Haridwar. Concentration of dissolved oxygen is one of the most important parameter to indicate water purity and to determine the distribution and abundance of various algal groups. It has been recommended that a minimum of 4 mg/l of dissolved oxygen should be maintained in water for healthy growth of fish and other microbial population. Different workers have pointed out various influencing factors on oxygen level that include discharges from industries, water current, velocity and biota.

Free carbon dioxide was observed maximum (4.80 mg/l ± 0.00) in the month of June due to higher turbidity, Velocity and water temperature, but was recorded minimum (1.13 mg/l ± 0.00) in February month due to low turbidity, velocity and water temperature. The turbulence velocity low in October to March, because of that the CO_2 atmospheric diffusion was also low causes further lowering of CO_2 concentration in water. Annual average value of free carbon dioxide varied between 2.33 mg/l ± 0.05 at sampling station E in 2003-2004 to 2.89 mg/l ± 0.11 in which maximum average value obtained from sampling point C in the year 2003-2004, Pahwa and Mehrotra (1966) and Ray *et al.* (1966) have reported that the Ganga river contains maximum free carbon dioxide in rainy season at Allahabad. Khanna *et al.* (1997) and Seth *et al.* (2000) have also reported the same trends in the river Ganga at Haridwar but Khanna and Singh (2000) observed maximum free carbon dioxide during summer in Suswa River at

Raiwala, Dehradun. The free carbon dioxide is released during the decomposition of certain substances and metabolic activity of the living organisms. Since higher temperature accelerate the decomposition of organic substances as well as the respiratory activity of the biota, a direct relationship was established between the water temperatures and free carbon dioxide.

The dissolved oxygen and free carbon dioxide usually inversely related to one another because of photosynthetic and respiratory activity of the biota (Hynes, 1970). The fluctuations in the dissolved oxygen content were mainly influenced by the factors like TOC, plankton and microbial activity.

Many studies also reveal that the alkalinity may also increase due to decomposition of organic matter that has settled at the bottom. The total alkalinity was maximum (825.00 mg/l ± 0.09) in the month of May and minimum (36.00 mg/l ± 2.40) in August. In the period of lower rate of precipitation accompanied with warmer and humid climatic condition, the degradation of organic matter is enhanced. Whereas during the Winter season the organic matter at the bottom is difficult to degrade because of low temperature of water. The annual average value of alkalinity varied between 63.37 mg/l ± 1.40 to 138.53 mg/l ± 5.53 in which maximum average value is obtained from sampling station D in the year 2003-2004 and minimum from sampling station A in the year 2004-2005. Similar observation were also obtained by Holden and Green (1960), Talling and Rzoska (1967) and Abdin (1948a,b). The decomposition of the organic matter leads to the high alkalinity of the waters. (Venkateswarlu & Jayanti, 1968).

The bio-chemical oxygen demand was observed maximum (5.64 mg/l ± 0.21) in July month and minimum (1.17 mg/l ± 0.00) in February month. The annual average value of biochemical oxygen demand ranged between 2.37 mg/l ± 0.14 to 3.33 mg/l ± 0.15. The minimum average value was found at sampling station E in the year 2004-2005 and maximum in the year 2003-2004 at sampling station E. Highest annual average value of bio-chemical oxygen demand at sampling station E may be due to drainage of several small sewage drains into the river and runoff of sludgy, silted sewage during months of rainy season. A negative relationship has been observed between BOD and DO contents. A similar pattern has been reported by Khanna (1993) and Prasad *et al.* (2003). BOD determination is a most useful technique to assess the level of organic pollution in river system. BOD levels were probably influenced by heavy metals with regard to seasonal fluctuation. Raina *et al.* (1984) reported peak values during winter in river Jhelum and Bhargava (1985) during summer for the river Yamuna. Khanna & Singh (2000) noticed peak values during summer in Suswa river and Khanna *et al.* (1997) observed peak values in monsoon season in river Ganga.

Chemical Oxygen Demand (COD) represents chemically oxidisable load of organic matter in water. It was noted highest (44.00mg/l ± 0.26) in the month of June and minimum (2.98 mg/l ± 0.62) in the month July. The annual average value of COD ranged between 7.26 mg/l ± 1.19 to 11.86 mg/l ± 0.77 whereas least average value was found at sampling station B in the year 2004-2005 and maximum at sampling station B in 2003-2004. Similar trends of COD have shown by Khanna *et al.* (2002) in the river Ganga and Khanna & Singh (2000) in Suswa River at Raiwala.

Maximum values of total hardness were recorded in August (406.00 mg/l ± 0.27) and minimum was recorded in March (27.50 mg/l ± 0.00). The annual average values of total hardness ranged from 59.72 mg/l ± 0.27 to 93.96 mg/l ± 0.27 in which maximum average value is recorded from sampling station A in 2003-2004 and minimum in 2004-2005 at sampling station A. Calcium and Magnesium are the two component of Total hardness. It is present in form of carbonate and bicarbonate. Calcium iron is one of the most abundant substances of natural waters. Being present in high quantities in the rocks, it is leached from there to contaminate the water. Calcium is essential for metabolic processes in all-living organisms. In this study the Calcium found in range of 10.21 mg/l ± 0.03 in December to 48.00 mg/l ± 0.01 in January. Maximum values of Magnesium were recorded in August (1859.32 mg/l ± 1.27) and the minimum value was observed in month of February (13.53 mg/l ± 10.07). Khanna *et al.* (1993, 2003) and Mishra & Joshi (2003) observed hardness in river Ganga at Haridwar and found more or less similar trends in their study.

The highest and lowest values of acidity were found in January and July month, 89.55 mg/l ± 2.52 and 18.25 mg/l ± 0.09 respectively. The annual average value of acidity varied between 46.12 mg/l ± 0.10 to 61.01 mg/l ± 1.03 in which maximum average value was recorded from the station A in the year 2004-2005 and minimum was at sampling station E in the year 2004-2005. Similar trend was found by Khanna and Bhutiani (2003).

The maximum value of chloride were found in the month of July (28.00 mg/l ± 0.00) and minimum in January (5.71 mg/l ± 6.13) respectively. The annual average value of chloride varied between 12.74 mg/l ± 0.39 to 19.87 mg/l ± 0.10 in which maximum average value was recorded from the station E in the year 2004-2005 and minimum was at sampling station D in the year 2004-2005 Chlorides are present in sewage, sewage effluents and farm drainage. Significant levels of chloride were shown by many rivers like Yamuna (Sengar *et al.*, 1885); Tungbhadra (Reddy and Venkateswarlu, 1987); Jhelum (Raina *et al.*, 1984) and Kshipra (Mishra and Saxena, 1984). CPCB (2003) reported the value of chloride in between 14 to 51 mg/l during Ganga monitoring from Bithur, Kanpur to Sangam, Allahabad. Mishra and Joshi (2003) reported chloride in Ganga river at Haridwar. CPCB (1990-91) also studied the chloride from Rishikesh to Haridwar and showed a significant increasing trend on chloride at all monitoring stations in West Bengal stretches.

The amount of nitrogen available in fresh water is less but they are of significant importance to the water ecosystem. In the majority of natural aerobic water most nitrogen occurs as nitrate but is more in eutrophic water. Nitrate was minimum (0.10mg/l ± 0.00) in the month of November and was maximum (2.87 mg/l ± 0.32) in the April month. This may be because nitrate is formed due to degradation of organic matter. The higher temperature enhances the degradation process. This is in confirmation to findings of Verma (1978) they had observed low values of nitrate in Winters and Rainy and more in Summer season. Khanna (1993) also found high nitrogen in the month of June. Where as Venkateshwarlu and Jayanti (1968) observed high concentration of nitrates after onset of Monsoon.

In the study of heavy metals Copper and Iron were taken for observation. The maximum value of copper in the month of August was 0.68 mg/l ± 0.16 and the minimum value was 0.10 mg/l ± 0.00 in the month of November. The annual average value of Copper ranged between 0.03 mg/l ± 0.03 to 0.21 mg/l ± 0.04. The minimum value was found from sampling station B in 2004-2005 and maximum from sampling station D in 2004-2005. The annual average value of Iron varied between 2.08 mg/l ± 0.02 to 0.31 mg/l ± 0.23 in which maximum average value of Iron was obtained at sampling station A in 2003-2004 and minimum average value at sampling station A in 2004-2005. Heavy metals get contaminated into aquatic systems as a result of various natural activities (Weathering of soils and rocks from volcanic eruptions) and from a variety of human activities involving the mining, processing or use of metals or substances. Some metals, such as copper and Iron are essential micronutrients virtually all metals, including the essential metal micronutrients, are toxic if exposure levels are sufficient high. The increased circulation of toxic metals in recent times resulted in the unavoidable build up of such toxic substances in the human food chain. Since heavy metals are rapidly absorbed to particulate materials (eg. detritus, plankton, suspended sediments) and assimilated by living organisms. Metals adhering particulate materials could be absorbed by an organism in the process of pumping water over its gills or in the process of filtering water to obtain food. There is paucity of information on the toxicity of suspended metals to most aquatic organisms. Khanna *et al.* (2003) also reported heavy metals in water of Ganga river at Bulandshahar.

The phytoplankton of many Indian rivers consists of diverse assemblage of major taxonomic groups. Many of these forms have different environmental and physiological requirements. The number, type and distribution of these organisms present in any aquatic habitat provide a clue on the environmental conditions prevailing in that particular habitat. It is seen that many environmental factors interact to provide conditions for the growth of plankton both specially and seasonally. Khanna (1993) emphasized the need for monitoring of a river continuously at least for ten years to assess the importance of various chemical parameters in the growth and distribution of plankton. Ray (1955) and Venkateswarlu and Menon (1969) recorded maximum values of total plankton during winter and minimum during rainy season. Lakshminarayana (1965), Pahwa and Mehrotra (1966) observed seasonal maximum in river Ganga both in summer and winter.

In the present work it has been noted that the maximum total phytoplankton were 8933.93 unit/l ± 300.62 were found in the month of January and zooplankton were 856.62 unit/l ± 132.00 in the month of March. The minimum concentration of total phytoplankton was recorded 362.28 unit/l ± 65.76 in the month of November and zooplankton were recorded 15.61 unit/l ± 12.42 in the month of July. The annual average value of zooplankton ranged between 257.69 unit/l ± 87.93 to 213.40 unit/l ± 42.00 in which maximum average value was recorded from sampling station D in the year 2003-2004 and minimum was recorded from sampling station D in the year 2004-2005. The annual average value of phytoplankton ranged between 1275.57 unit/l ± 185.88 to 3454.49 unit/l ± 342.80 in which maximum average value was obtained from sampling station E in 2003-2004 and minimum was obtained from sampling station A in the year 2004-2005. The study showed a dominance of phytoplankton as also reported by Pahwa and Mehrotra (1966) in the Ganga river. Badola and Singh (1981), Dobriyal and Singh (1987), Khanna (1993) and Khanna *et al.* (2000) observed high values of plankton during January-March. Khanna and Singh (2000) also reported high values of plankton during winter season.

The Planktonic concentration also affected by the turbidity. The increased turbidity reduces the plankton production (Khanna *et al.*, 1993). In the present study it was noted that temperature show a negative relationship with plankton. The plankton was maximum in November to March when the temperature was low. Eddy (1934) and Chandler (1940) pointed out that the plankton production is mainly influenced by temperature. Khanna (1993), Khanna *et al.* (2000) also reported the same trends. The results indicated that the plankton were maximum in the winter month probably due to low temperature, high content of DO and low velocity. Similar study was made by Khanna and Bhutiani (2003) and Khanna *et al.* (2000).

The benthic fauna consists of minute organisms that move around in water. These constitute the main bulk of food of bottom feeding fishes and are relatively less exposed to environmental fluctuations than the organisms on the surface water, thus they serve as bioindicators of ecological conditions of a habitat. Singh *et al.* (1994) also performed a qualitative and quantitative study of benthic fauna of Kararia Lake. As per Singh, Singh and Singh (2002) these are of great importance for biological monitoring. These serve as food for fish community, along with plankton. Thus they serve as an important section of water ecology. Although the population of benthic fauna has shown no significant. Monthly variation from one point of sampling to another but the annual total observation showed that *Neuroptera* was the most dominant benthic fauna found to be (48.25 Units/m^2) in 2003-2004 at sampling station A, and *Turbellaria* shown the minimum value of 0.92 Units/m^2 in 2004-2005 at sampling station E. In Ganga water, the production of benthic fauna is mainly influenced by temperature and light. Jindal and Singh (2005) had similar observations in a study of fresh water ecosystem. Temperature and sunlight have also a major contribution in productivity and growth of the alga, hydrophytes and other phytoplankton. Although the temperature of river is low in Winters even then due to increased water transparency the solar rays and energy can be transmitted more, which is a reason for the flourishing phytoplankton and hydrophytes in winters (Khanna 1993). These may serve as support for the life of benthic fauna, providing food, shelter and also required DO which increases in water due to higher rate of photosynthesis in Winter season. So, it can be accounted as one of the major factors for the rich diversity of fish fauna in the river Ganga.

In the present study the benthic fauna were more in December and January and slightly less in May to September. This can be due to increased velocity of flooded river which swipes away most of the phytoplankton, macro-invertebrates, nymphs, insect larvae and diatoms. Furthermore, the turbidity in river is high in Monsoon. This finding is similar to that of Khanna (1993). Prasad and Singh (2003) observed that as the effect of increased volume and velocity of water plankton and benthic fauna were disturbed and get decreased. In some places Annelids and Gastropods were observed during Monsoon. Higher suspended solids serve as a blanket, reduces the penetration of solar light and energy required for growth of hydrophytes and phytoplankton, this interferes with photosynthesis activity and also reduces the oxygen concentration in water therefore the Benthic fauna gets reduced. In Winter season, the situation is however, vice versa as because of low turbidity sunrays can penetrate deeper in water and increase photosynthetic activity. This was also reported by Bisht and Das (1985). Phytoplankton has however been extensively used till date, as indicators of water quality Shrivastava *et al.* (1996) made observations on productivity of phytoplankton in relation to physico-chemical properties of Rapti river, Somashekhar (1984) also employed phytoplankton as bioindicators. Khanna (1993) made note on the plankton ecology of river Ganga at Chandighat, Haridwar. Khanna *et al.* (1997) made efforts to relate physico-chemical factors of the river Ganga at Lal-ji-wala with population of green algae, Haridwar. Shukla and Srivastava (2003) made attempt to relate the pollution status of Gandhisagar reservoir with macro-zoo benthos saprobiety index.

The present study reveal that the value of Most Probable Number (MPN) coliform bacteria increases from summer season, maximum value was noted (27000.00 MPN/100ml ± 80.00) in the month of July and minimum was observed (25.00 MPN/100 ml. ± 30.00) in the month of May. The annual average values of total coliform varied between 1543.08 MPN/100 ml. ± 50.42 to 4908.33 MPN/100 ml. ± 36.00. Average value of total coliform was found maximum at sampling station E in the year 2004-2005 and minimum at sampling station D in 2003-2004. Average highest values of total coliform may be due to drainaging of several sewage drains, increasing human activities during summer. Causes of average minimum value of total coliform may be due to river stream not flowing, touching the town situated at this point and no sewage contaminations are found at this point. Singh *et al.* (1988; 1989) reported the coliform bacteria maximum 16000 MPN/100 ml and minimum 940 MPN/100ml before Kumbh mela in 1988 in Ganga, Yamuna rivers and Sangam, at Allahabad. In 1989 during Kumbh mela MPN values reached upto 130000 MPN/100 ml at Sangam and after Kumbh mela maximum MPN value was found upto 9200 MPN/100 ml. at Sangam. Khanna and Chugh (2004) in the river Ganga at Haridwar reported the similar trends of total coliform bacteria.

The higher value of Standard Plate Count (SPC) was recorded in the month of July (262.00 × 10^3 SPC/ml ± 2.50) and minimum in the month of February (7.00 × 10^3 SPC/ml. ± 1.00). Average annual value of SPC varied from 61.63 × 10^3 SPC/ml ± 1.66 to 108.29 × 10^3 SPC/ml ± 3.06 in which maximum average value of SPC was recorded at sampling station C in the year 2004-2005 and minimum was recorded at sampling station E in 2003-2004. Similar trends of SPC was also studied by Khanna & Chugh (2004) in Ganga River at Haridwar. Kapadnis and Panse (1988) studied the microbial population in water of swimming pools in the city of Pune and reported total viable count from100 to 3200 cfu/ml. It is also noted that all sampling stations are at near by of any village or city, the *Escherichia coli, Streptococcus facecalis*, *Staphylococcus aureus* and *Clostridium perfringes*, were found due to the mixing of domestic and city sewage in the main stream. Bonde (1977) reported *Escherichia coli,* as bacterial indicator of fresh water pollution. He also showed Salmonella and other pathogenic bacteria in fresh water. Dutta and Bell (1973) and Borego *et al.* (1990) reported coliphages as an indicator of pollution.

CHAPTER-6

SUMMARY

Of all the earth ecosystems, rivers are the most dynamic having as their primary function to transport of water. Rivers have been studied for many years with inputs from many discipline of sciences and humanities. The stretch of Ganga river has varied geological and climatic conditions. It is because of this reason that it attracts not only the tourists and pilgrims from all over the world but also many research workers to make untiring efforts to investigate and discover the lotic environment of the Mother river of India that gives shelter to a rich biodiversity. The Ganga basin is one of the richest river basins. The water of Ganga river has traditionally been regarded as inexhaustible gift of nature. The natural environmental composition in this study stretch is facing fast changes because of growing population and human habitat, the unjudged exploitation of the river, mass bathing during festive occasions, dumping of wastes and effluents, use of chemicals for nearby irrigational lands, rapid industrialization for development.

However, these and many more put severe strains on the Ganga and to an extent resulted in some degradation in its quality. The problem of the degrading water of the river has not just come up but is on the way to take serious dimensions in certain stretch.

The summary of the findings is mentioned below:

1. The difference between water temperatures of the river was lesser in the months of October to April in comparison to April to October. The predicted reason for this may be the melting snow from the peaks of Himalayas during peak Summer that resists the major change in temperature during the two seasons. In the present study, a difference in the fluctuation of water temperature was observed 10.00°C ± 0.00 (minimum) in the month of November and 25.75°C ± 1.07 (maximum) in March. Annual average values of temperature varied between 14.55°C ± 1.78 to 17.55°C ± 0.44.
2. The conductivity ranged between minimum 32.11 Simens/cm ± 0.24 to maximum of 360.07 Simens/cm ± 0.00. It can be said that the present higher conductivity values in August to October month due to the input of large amount of salts and silt carried by the river.
3. The water of the river was generally clean in other season with lowest turbidity in Winter season. In the month of February and March some sampling station recorded Nil turbidity and highest (1264.00 JTU ± 13.752) noted in the month of September.
4. The Maximum velocity (3.90m/sec. ± 0.75) recorded in July month and minimum of 0.10 m/sec.± 0.00 in the month of March.
5. In the present investigation it was noted that the total solids were maximum in the month of June (1793.68 mg/l ± 0.16), which were responsible for the turbidity in the river. The total solids were recorded minimum (1.50 mg/l ± 0.82) in the month of February due to gradual sedimentation of the filterable residue.
6. In our study T.D.S. found minimum of 62.15 mg/l ± 0.20 in the month of March and the maximum upto 895.65 mg/l ± 2.54 in the month of July, with the annual average ranges between 122.80 mg/l ± 12.29 to 270.82 mg/l ± 3.97.

7. Total suspended solid having the maximum and minimum value range observed during out study was 1476.08 mg/l ± 0.10 in the month of June to 1.90 mg/l ± 60.86 in November.

8. The pH was maximum (8.90± 0.07) in November at sampling station E may be due to increased ashes from near by ashrams and washing activities at Harkipauri. The pH was recorded to be minimum (5.94) in the month of February at sampling station E.

9. Maximum dissolved oxygen was recorded (14.35 mg/l ± 0.00) in the month of December at the sampling station E. The minimum value of dissolved oxygen (6.17 mg/l ± 0.12) was observed in the month of June at sampling station E.

10. Free carbon dioxide was observed maximum (4.80 mg/l ± 0.00) in the month of June due to higher turbidity, Velocity and water temperature, but was recorded minimum (1.13 mg/l ± 0.00) in February month due to low turbidity, velocity and water temperature. The turbulence velocity low in October to March, because of that the CO_2 atmospheric diffusion was also low causes further lowering of CO_2 concentration in water.

11. The total alkalinity was maximum (825.00 mg/l ± 0.09) in the month of May and minimum (36.00 mg/l ± 2.40) in August.

12. The estimation of biochemical oxygen demand (BOD) was done to assess the organic load in river system. The biochemical oxygen demand was observed maximum (5.64 mg/l ± 0.21) in July month and minimum (1.17 mg/l ± 0.00) in February month.

13. It was noted highest (44.00mg/l ± 0.26) in the month of June and minimum (2.98 mg/l ± 0.62) in the month July. Both, biochemical oxygen demand (BOD) and chemical oxygen demand (COD) showed an inverse relation with dissolved oxygen (DO).

14. Maximum values of total hardness were recorded in August (406.00 mg/l ± 0.27) and minimum was recorded in March

(27.50 mg/l ± 0.00). The annual average values of total hardness ranged from 59.72 mg/l ± 0.27 to 93.96 mg/l ± 0.27.

15. In this study the Calcium found in range of 10.21 mg/l ± 0.03 in December to 48.00 mg/l ± 0.01 in January.
16. Maximum values of Magnesium were recorded in August (1859.32 mg/l ± 1.27) and the minimum value was observed in month of February (13.53 mg/l ± 10.07).
17. The maximum value of chloride were found in the month of July (28.00 mg/l ± 0.00) and minimum in January (5.71 mg/l ± 6.13) respectively. The annual average value of chloride varied between 12.74 mg/l ± 0.39 to 19.87 mg/l ± 0.10.
18. Nitrate was minimum (0.10mg/lt ± 0.00) in the month of November and was maximum (2.87 mg/l ± 0.32) in the April month. This may be because nitrate is formed due to degradation of organic matter.
19. The maximum value of copper in the month of August was 0.68 mg/l ± 0.16 and the minimum value was 0.10 mg/l ± 0.00 in the month of November. The annual average of Copper ranged between 0.03 mg/l ± 0.03 to 0.21 mg/l ± 0.04.
20. The annual average value of Iron varied between 2.08 mg/l ± 0.02 to 0.31 mg/l ± 0.23.
21. In the present work it has been noted that the maximum total phytoplankton were 8933.93 unit/l ± 300.62 were found in the month of January and zooplankton were 856.62 unit/l ± 132.00 in the month of March. The minimum concentration of total phytoplankton was recorded 362.28 unit/l ± 65.76 in the month of November and zooplankton were recorded 15.61 unit/l ± 12.42 in the month of July.
22. The annual total observation showed that *Neuroptera* was the most dominant benthic fauna found to be (48.25 Units/m^2) in 2003-2004 at sampling station A, and *Turbellaria* shown the minimum value of 0.92 Units/m^2 in 2004-2005 at sampling station E.

23. The present study reveal that the value of Most Probable Number (MPN) coliform bacteria increases from summer season, maximum value was noted (27000.00 MPN/100ml ± 80.00) in the month of July and minimum was observed (25.00 MPN/100 ml. ± 30.00) in the month of May.

24. The higher value of Standard Plate Count (SPC) was recorded in the month of July (262.00×10^3 SPC/ml ± 2.50) and minimum in the month of February (7.00×10^3 SPC/ml. ± 1.00).

REFERENCES

Allan, J. D. (1995). Stream Ecology : Structure and Functioning of Running Water, *Chapman and Hall*, London.

Allan, J.D. and Flecker, A.S. (1993). Biodiversity and conservation of running water. *Bioscience*, 43: 32-43.

Abdin, G (1948a). Physical and chemical investigations relating to algal growth in river Nile. *Cairo. Bull. Inst. Egypt*, 29:20-24.

Abdin, G (1948b). Seasonal distribution of Phytoplankton and sessile algae in the river Nile. *Cairo. Bull. Inst. Egypt*, 29:369-382.

Abraham, Beena T., Anirudhan, T.S. and Sumesh Kumar, K.S. (2002). Distribution of hydrographic parameters and heavy metal ions in river Periyar. *Nature Environment and Pollution Technology,* 1 (2), pp 161-163.

Abramovitz, J.N. (1995). Freshwater Failures: the crisis on five components. *World Watch*, 8: 26-28.

Agarwal, D.K., Gaur, S.D., Tiwari, I.C., Narayanaswami, M.S. and Harwali, S.M. (1976). Physico-chemical characteristic of Ganga at Varanasi. *Indian J. Env. Health*, 18: 201-206.

Aggarwal (1986). Ecology of Chambal River at Kota. *Acta Ecologia,* 8(1):13-19.

Ahmed, N. (1952). Stocking of fish in East Pakistan. *Agric. Pakist.*, 3 (4):1-3.

Ahmed, K and Begum, J. (2002). Bacteriological quality of water in hill stream of Khanpara during different seasons. *Nature Environment and Pollution Technology,* 1(2), 171-173.

Alabaster, J.S. and Lloyd, R. (1980). *Water quality criteria for fresh water fishes*, Butter worth, London.

APHA, AWWA and WPCF (1995). *Standard methods for the examination of water and wastewater*. American Public Health Association. Washington 19th Edition. N.York pp.1750.

Badola, S.P. and Singh, H.R. (1981). Fish and fisheries of river Alaknanda *Proc. Nat. Acad. Sci.,* 15(B):133-142.

Badola, S.P. and Pant, M.C. (1973). Fish fauna of the Garhwal Hills part -I. Ind. J. Zool., 14 (1): 37-44.

Badola, S.P. (1979). Limnological studies of Icthyofauna of some fresh water resources of Garhwal region. *D.Phil. Thesis*, Garhwal University.

Badola, S.P. and Singh, H.R. (1981). Hyrdrobiology of river Alaknanda of Garhwal, 8 (2):269-276.

Badola, S.P. and Badola, S. (1999). *Observations on destructive fishing methods and recommendations to protect fish fauna in Garhwal Himalayas*. Sus. Eco. and Env., Edited by Gautam Ashutosh, Khanna, D.R. and Gautam Aditya. pp.165-168.

Bhargava, D.S. (1985). Water quality variations and control technology of Yamuna River. *Environ. Pollut.* (Series A), 37:355-372.

Bahuguna, S.N. and Badoni, A.K.(2002). Qualitative and quantitative productivity of fish food from three important tributaries of river Alaknanda, Garhwal. *Him. J. Env. Zool.,* Vol.16(2), pp.215-222.

Bahumik, U. and Saha, S.K. (1998). Role of extension in arising mass awareness and public participation in fish conservation movement. In Ponniah, A.G., Das, P. and Verma, S.R. (Eds.) *Fish Genetics and Biodiversity Conservation*. Natcon Pub. No.-5, Muzaffarnagar pp. 157-163.

Basu, M. and Mitra, A.K. (2002). studies on the pollution of river Damodar at Burdwan with special reference to metals, COD and pH. *Nature Environment and Pollution Technology*, 1 (4), pp 397-399.

Berner, L.M. (1951). Limnology of lower Missouri river, *Ecology*, 32: 1-12.

Bharti, A., Saxena, R.P. and Pandey, G.N. (1979). Physiological imbalances due to Hexavalent Chromium on fresh water algae *Indian. J. Environ. Health*, 2(3): 234-243.

Bhatt, S.D., Bisht, Y. and Negi, U. (1984). Ecology of Limnifarm in river Kosi of the Kumaun Himalaya (U.P.). *Proc. Indian Nalu. S.C. Acad. B.*, 50(4) : 395-405.

Bhatt, S.D., Bisht, Y. and Negi, U. (1985). Ecology of the limnoflora in river Kosi of Kumayon, Himalaya (U.P. Proc. Natl. Sci. 50 (4):395-405.

Bhattacharya, T. and Saha, R.K. (1988). A comparative study of the physico-chemical properties and plankton density of two streams and a tributary of Gomti river in Tripura. *Geobios*. Vol. 15(5) : 201-206.

Bilgrami, K.S. and Duttamunshi, J.S. (1985). Ecology of river Ganges (Patna-Farakka). Technical Report, CSIR.

Bilgrami, K.S. and Duttamunshi, J.S. (1979). Limnological survey and impact of human activities on river Ganges (Barauni – Farrakka). Technical Report Submitted to D.S.T. p.91.

Bisht, R.S. and Das S.M. (1985). Abiotic parameters and aquatic insects of two cold water Kumaon lakes. *Indian J. Ecol*, 12 (2) : 365-369.

Bissonette, G.K., Jezeski, J.J., Mc. Feters, G.A. and Steuart (1975). Influence of environmental stress on the enumeration of indicator bacteria from natural waters. *Appl. Microbiology*. 29:186-194.

Bowaakar, A.P. (1976). The physico-chemical limnology of Mwenda Mouth Lake, Koria. *Arch. Hydrobiol.*, 77 (1):60-108.

Bonde (1977). Bacterial indication of water pollution. In: Droop, M.R. and Jannash, H.W. (Eds) *Advances in microbiology*, Vol.1, Acad. Press. London.

Borego, J.J., Cornax, R., Morningo, M.A., Manzanares, E.M. and Romero, P. (1990). Cephalophages as an indicator of fecal pollution in water. Their survival and productive infectivity in natural aquatic environments. *Water Res.*, 24 (1) :111-116.

Chako, P.I. and Ganpati, S.V. (1949). Some observations on the Adyar river with special reference to its hydrobiological conditions *Indian Geogr. J.,* 24 (3).

Chako, P.I. and Ganpati, S.V. (1952). Hydrobiological survey of the Surli river in the highway of the Madurai district, India, to determine its suitability for the introduction of rainbow trout. *Arch. Hydrobiol.*, 48:128-141.

Chattopadhya, S.N., Routh, T., Sharma, V.P., Arora, H.C. and Gupta, R.K. (1984). A short term study on the pollution status of river Ganga at Kanpur region. *Indian J. Env. Health*, 26(3): 244-257.

Chandler, D.S. (1940). Limnological studies of western Eric. I : Plankton and certain physico-chemical Data of Bass Island region from September 1938 to November, 1939. Ohio, *J. Sci.*, 40 : 291-336.

Chaphekar, S.B. and Mahatre G.N. (1983). A report on the impact of human settlement and development activities on Ganga river system. *Inst. of Sci. Bombay.* pp.65.

Chakrabarty, R.D., Ray. P. and Singh, S.B. (1959). A quantitative study of plankton and physiological condition of river Jamuna at Allahabad in 1954-1955. *Indian J. Fish,* 6 (1): 186-203.

Chopra, A.K., Madhwal, B.P. and Singh, H.R. (1990). Abiotic variable and Prima productivity of river Yamuna at Naugaon, Uttar Kashi, Garhwal. *Indian J. Ecol.* 17(1) : 61-64.

Chopra, A.K. and Patrik J.N., Nirmal J. (1994). Effect of domestic sewage on self-putrification of Ganga water at Rishikesh. Physico-chemical parameters. *Ad. Bids.* Vol. 13(1): 75-82.

Chopra, A.K. and Arya, J.S.(1996). Physico-chemical and microbiological characterization of waste water effluents of different sites of thermal power station, BHEL, Haridwar. *Him. J. Env. Zool., Vol.* 10, pp.35-37.

Chopra, A.K. and Rehman, A. (1995). A study on self-purification of physico-chemical properties of Ganga canal water at Jwalapur, Haridwar *Him. J. Env. Zool.*, Vol.9: 11-13.

Chugunova, N.I. (1963). Age and growth in fish. National Science Foundation, Washington pp. 132.

C.P.C.B. (1990-91). Annual Report. Central Pollution Control Board, Parivesh Bhawan, East Arjun Nagar, Delhi. pp. 7-21.

C.P.C.B. (2003). A report on colour problem of river Ganga, Central pollution Control Board, Zonal Kanpur. p.1-9.

Cook, J. (1977). Environmental pollution by heavy metals. Internet, Environmental Studies 9: 253-256.

Cosy, H. and Newton, P.V.R. (1973). The chemical composition and flow of river Frome and its main tributaries. *Fresh Water Biol.* 3: 317-333.

Dabron, G.R. (1976). Physical and chemical factors of vernal temporary pond in western Canada. *Hydrobiologia* SI (1): 33-39.

Daniel,C.C., Wilder, H.B. and Weiner, M.S. (1982). Water quality of French Broad river, North California. An analysis of data collection at Marshall between 1958-77. *Wat.Supp.Pap. 2185 A-D US Geol. Surv.*, Govt. Print. Press, Washington, D.C.

Das, P. (1994). Strategies for conservation of threatened fisheries. In Dehadrai, P.V., Das, P and Verma, S.R. (Eds.) *Threatened Fishes of India*. Pub.No. 4, Muzaffarnagar, pp. 307-310.

Das, S.K. and Tiwari, N.P. (1996). Seasonal variation in the productivity of some macro invertebrates population of Bagulia Lake, Vidyapati Nagar, Bihar. *Him. J. Env. Zool.,* Vol. 10,67-69.

Das, S.M. and Moitra, S.K. (1955). Studies on the food of some common fishes of Uttar Pradesh, India. The surface feeders, the mid feeders and the bottom feeders *Proc. Nat. Acad. Sci. India,* 25 B (1 and 2): 1-6.

Das, S.M. and Akhtar, S. (1970). A report on fresh water plankton from Dul Lake, *Kashmir. Sc.*, 7 (1-2): 133-137.

Das, H.B., Kalita, H., Saikia, L.B., Borah, K. and Kannur, S. (1992). Physico-chemical characteristics of Brahmaputra water of Tezpur. *Poll. Res.* 11 : 169-172.

Deming, H.G. (1975). Water the fundamental opportunity.Oxford University Press, NewYork. Pp. 942.

Deshmukh, S.B., Phadke, N.S. and Kothandaraman, V. (1964). Physico-chemical study of Kanhan River water (Nagpur) *Indian J. Environ. Health,* 6(3): 181-186.

Dwivedi, H.N. Pandey, B.N., Kumar, M. and Alam, Naushad (1995). Limnological study in relation to fish production in two ponds at Patna. *Him. J. Env. Zool.* , Vol.9, pp.7-10.

Dobriyal, A.K. and Singh,H.R. (1981). Diurnal variation in some aspects of limnology of the river Mandakani from the Garhwal Himalaya. *Uttar Pradesh J. Zoo.,* 1:16-18.

Dobriyal, A.K., Singh, H.R. and Bist, K.L. (1983). Diurnal variation in hydrobiological parameters of two hill streams of Garhwal Himalaya. *Uttar Pradesh J. Zoo.,* 3:30-34.

Dobriyal, A.K. (1985). Ecology of limnofauna in the small streams and their importance to the village life in Garhwal Himalaya. *Uttar Pradesh J. Zoo.,* 5(2):139-144.

Dobriyal, A.K. and Singh, H.R. (1987). A case study on the origin of rhithroplankton in Garhwal Hill streams. *Agri-Biol.* 3(2) : 104-106.

Dobriyal, A.K. and Singh, H.R. (1988). Ecological basis for ichthyofaunal variation in two hill streams of Garhwal Himalaya. *Proc. First Indian Fisheries Forum*, Manglore, 1987.

Dua, A. and Johal, M.S. (2000). Use of fish scales as an indicator of pesticide pollution. *Proceeding National Symposium on 'Aquaculture for 2000 A.D.'* Organized by Madurai University, Madurai, India. PP. 283-288.

Dua, A. and Johal, M.S. (1994). Alterations in the architecture of gill surface of *Channa pinctatus* produced by endosulfan treated water: A SEM study. *J. Aqu. Biol. Fish.* Vol. (1): 27-31.

Dutta, B.J. and Bell, J.B. (1973). Isolation of Salmonella from moderately waters. *J. Wat. Pollut. Control Fed.*, 45, 316-323.

Dutta, S.P.S., Malhotra, Y.R., Sharma, K.K., and Sinha, K. (1997). Diel variations in physico-chemical characteristics of water and macrobenthic invertebrates in river Tawi, near Nagrota by pass, Jammu. Him. *J. Env. Zool.*, Vol.11, pp.45-50.

Dutta, A., Ahmad,T.Md.,Mukherjee,M., Dutta, S. and Pandey, P. (2002). Presented an overview of the fish diversity of Ranchi district *Him. J. Env. Zool.* , Vol.16(1), pp.257-262.

Eddy, S. (1934). A study of the fresh water plankton communities. *Illinois, Biol.* Monyar, 12(4) : 1-93.

Esmaeili, H.R., Johal, M.S., Tandon, K.K. and Sharma, M.L. (2001). SEM studied on body scale of a tooth carp, *Aphanius ginaonis* (Holly) from the hot spring of Geno, South Iran. *Abstract Book , XXXIV Annual Conference of EMSI on Electron Microscopy and Allied Fields*. pp.42-43.

Forel, F.A. (1892). Le Leman. *Monographic Limnogique*, 3 Vol. Lavsane.

Fokmare, A.K. and Musaddique, Mohmmad (2002). A study on the physico-chemical characteristics of Kapashi lake and Purna river waters in Alok district of Maharashtra, India. *Nature Environment and Pollution Technology,* 1(3), pp. 261-263.

Gautam, A. (1990). *Ecology and pollution of mountain waters*, Ashish Pub. House. N. Delhi. pp.206.

Gautam, A. (1990). Sediments of the river Bhagirathi, Garhwal Himalaya. *Geobios new Report* 9: 84-87, 1990.

Gautam, A. (1992). Some considerations on stone removals from river basin and resultant effects. *J. Ecobiol.* 4 (1), 55-58.

Gautam, A (1995). Continuous monitoring is an effective way to control water pollution. *J. Env.and Poll.*, 2(4) pp 173-176.

Gautam, A., Singh, H.R. and Sati, O.P. (1989). Seasonal variation of certain oxidation- reduction characteristics of the river Bhagirathi. *Proc. Indian. Nat. Sci. Acad.,* B 55 No.2 pp. 111-114.

Guatam, A., Joshi, O.P. and Sati, O.P.(1993). Physico-chemical characteristics of sewage and its impact on water quality of Alaknanda at Srinagar (Garhwal). *J. Ecotoxicol. Environ. Monit.* 3 (1) 061-063.

Gautam, A., Sarkar, P. and Khanna, D.R. (2001). Quality deterioration in aquatic environment of the Ganges: A growing problem for future. *Indian J. Env. Sci.,* 5(2) pp.113-117.

Gautam, A., Khanna, D.R. and Sarkar, P. (2000). Diurnal variation in the physico-chemical characteristics of Ganga water at Triveni ghat, Rishikesh during winter season. *Indian J. Environ. & Ecoplan.,* 3 (2): 369-371.

Ghosh, B.B. and Basu, A.K. (1968). Observation on estuine pollution of the Hoogly by the effluents from a chemical factory complex at Rishra, W.B. (India) *Environ. Health,* 10:204-218.

Gopal, B. and Sah, M. (1993). Conservation and management of rivers in India; Case study of the Yamuna. *Environmental Conservation,* 20(2) : 243-253.

Govindan, V.S. and Sundaresan, B.B. (1979). Seasonal succession of algal flora in polluted region of Adyar river. *Indiar. J. Env. Health,* 21 (2):131-142.

Griffith, R.S. (1955). Analysis of plankton yields in relation to certain physical and chemical factors of lake Michigan. *Ecology*, 36 (4): 543-552.

Grove, A.T. (1972). The dissolved and suspended solids load carried by south west African rivers Senegal, Niger, Benue and Shoril. *J.Hydrol.* 10:277-300.

Grover,S.P., Aggarwal, B.S. and Rauthan, J.V.S. (1994). Icthyofauna of Doon Valley. *Him. J. Env. Zool.*, Vol.8, pp.128-133.

Goyal, S.M., Gebra, C.P. and Melnick, J.L. (1977). Occurrence and distribution of bacterial indicators and pathogens in canal communities along the Texas coast. *Appl. Environ. Micribiol.* 34: 139-149.

Hart, G.E. and Fuller, S.L.H. (1974). Pollution studies of fresh water invertebrates. New York Academic press., 230 pp.

Hora, S.L. (1921). On some new or rare species of fishes from Eastern Himalayas. *Rec. India Mus.*, 22/5: 731-744.

Holden, J.M. and Green, J. (1960). Hydrobiology and plankton of river Sokkoto. *J. Anim. Ecol.,* 29(1):65-84.

Holander, R.R. (1986). Microanalysis of scales of poecillid fishes. *Copeia* 1: 86-91.

Horned, D.A. (1982). Water quality of Neuse river North Carolina-variability, pollution loads and long-term trends. *Water supply paper 2185 A-1. U.S. Geol. Surv.,* Govt. Printing office, Washington D.C.

Hora, W.S. (1957). *Physiology of fishes*. Acad.Press Inc., New York. pp.640.

Hynes, H.B.N. (1970). *The biology of polluted waters Liver Poll.* The University press, pp:202.

Hutchinson, G.E. (1941). Ecological aspect of succession in natural population, *Amme. Nat.* 75- 406-418.

Jeejabai, N. and Rajendra, S. (1982). Phytoplankton composition as indicators of water quality-A study of Adyar river. *Proc. Muragappa Chettiar Symposium.*

Jena, B., Sudarshana, R. and Chaudhuary, S.B. (2003). Studies on water quality parameters around Sagar island, Sundarbands. *Nature Environment and Pollution Technology,* Vol.2 No.3 pp. 329-332.

Jhingran, V.G. (1975). *Fish and Fisheries of India*, Hindustan Publishing Corporation (India): 1-954.

Jhingran, V.G. (1991). *Fish and fisheries of India* (3rd Edition) , Hind. Pub. Corp., Delhi, India. pp.985.

Jhingran, V.G. and Sehgal, K.L. (1978). Cold water fisheries of India. *Central Int. Fish Res. Inst.*, Barrackpore, 1-232.

Jindal, R. and Singh, H. (2005). Ecology of benthos of some fresh water systems of Punjab. *Proceedings of National Seminar 'New trends in fish development in India',* Punjab University. pp. 179-184.

Johal, M.S., Armantrout, Neil.B. and Arunachalam, M.(2005). Changing Indian Rivers. *Proceedings of National Seminar 'New trends in fish development in India'*, Punjab University. pp. 35-41.

Johal, M.S. and Tyor, A.K. and Rawal, Y.K. (2005). Role of human beings vis-à-vis stream morphology an water quality of Western Himalayan streams. *Proceedings of National Seminar 'New trends in fish development in India'*, Punjab University. pp. 141-146.

John. V. (1976). Hydrobiological studies on the river Kallayi in Kerala. Indian *J. Fish.*, 23 : 72-85.

Joshi, B.D. (2005). On the ecology and contribution of smaller mountain tributaries to the productivity of major river Bhagirathi-Ganga- in Himalayan region. *Proceedings of National Seminar 'New trends in fish development in India'*, Punjab University. pp. 241-251.

Joshi, B.D., Bisht, R.C.S., Joshi, Namita and Singh, Rajendra (1991). A study of planktonic and benthic components of three selected tributaries of river Ganga between Devprayag and Rishikesh. *Him. J. Env. Zool.*, Vol. 10, pp. 23-26.

Joshi, B.D. and Pathak, J.K. (1991). A relative study of some physico-chemical parameters of sewage water at Uttarkashi. *Him. J. Env. Zool.*, Vol.5, pp. 53-56

Joshi, B.D., Pathank, J.K., Singh, Y.N., Bisht, R.C.S. and Joshi, P.C. (1993). Physico-chemical characteristic of river Bhagirathi in the uplands of Himalya. *Him. J. Env. Zool.*, 7: 64-75.

Joshi, B.D., Bisht, R.C.S. and Semval, V.P. (1995). Primary productivity in western Ganga canal at Haridwar. *Indian J. Ecol,.* 22 (2): 123-126.

Joshi, B.D., Bisht, R.C.S. and Joshi, Namita (1996). Planktonic population in relation to certain physico-chemical factors of Ganga canal at Jwalapur, Haridwar. *Him. J. Env. Zool.,* Vol.10 : 75-77.

Joshi, C.B. (1996). Hydro biological profile of river Sutlej in its middle stretch in western Himalaya. *Uttar Pradesh J.. Zool.,* 16 (2) : 9-103.

Joshi, P. (1994). Status of fish conservation in river Ramganga. In: Dehadrai, P.V., Das, P. and Verma, S.R. (Eds.) *Threatened fishes of India*. Natcon Pub. No. 4, Muzaffarnagar. pp. 149-153.

Joshi, S.N., Tripathi, G. and Tiwari, H.C. (1993). Fish and fisheries of Gori Ganga. In: Singh, H.R. (Eds.) *Advances in limnology*. Narendra Pub. House, N. Delhi. pp. 361-366.

Joshi, V.P., Gautam, A.and Sati, O.P. (1991). Diurnal variations in physico-chemical characteristics of Alaknanda river at Srinagar during winter season. *Geobios new Reports*, 10: 166-167.

Kapadnis, B.P. and Panse, M.V. (1988). Bacteriological analysis of water from swimming pools in Pune city (Maharashtra). Short communications. *Indian J. Env. Hlth.* 30(2) : 176-179.

Kaur, H., Bath, K.S., Mander, G. and Dhillon, S.S. (1999). Aquatic invertebrate diversity of Kalanji lake, Punjab. *Indian. J. Env. and Ecoplan.,* 2 (1) : 37-41.

Karuppasamy, R. (2000). Short and long term effect of Phenyl mercuric acetate on the protein metabolism in *Channa punctatus* (Bloch). *J. Natcon.* 12 (1): 83-93.

Kaur,S and Joshi, Namita (2002). The solid waste generation during normal days in Haridwar city. *Him. J. Env. Zool.,* Vol. 16 (2) , pp 267-270.

Kaur, Surendra and Joshi, B.D. (2003). Seasonal variation in some physico-chemical parameters of river Ganga in and around Haridwar. *Him. J. Env. Zool. ,* Vol 17 (1) pp. 45-55.

Khanna, D.R. (1993). *Ecology and pollution of river Ganga*. Ashish Pub. House, N.Delhi, 1-241.

Khannna, D.R. and Chugh, T. (2004). Microbial Ecology- A study of river Ganga Discovery Publishing House, N.Delhi, pp.1-277.

Khanna, D.R. and Badola, S.P. (1990). Icthyofauna of the river Ganga at foothills of Garhwal Himalayas. *Journal of Natural & Phy. Sci.,* Vol. 4, nos.1-2 pp.153- 162.

Khanna, D.R., Badola, S.P. and Dobriyal, A.K. (1993). Plankton ecology of river Ganga at Chandighat, Haridwar. *Advance in limnology*. Ed. By H.R. Singh, Narendra Publishing House, N.Delhi. 171-174.

Khanna, D.R. Malik, D.S. and Badola, S.P. (1997). Population of Green algae in relation to physico-chemical factors of the river Ganga at Lal-ji-wala, Haridwar. *Uttar Pradesh J. Zool.*, 17 (3) : 237-240.

Khanna, D.R., Malik, D.S.and Rupendra (1998). Fish fauna of the river Ganga at Rishikesh. *J. Natcon*, 10(2): 212-213.

Khanna, D.R., Gautam, A., Chug, T. and Sarkar, P. (2000). Impact of abiotic factors on the planktonic population of a pond. *Env. Cons. J.*, 1(1): 41-46.

Khanna, D.R., Gautam, A. and Sarkar, P. (2001). Water quality of Bathing Ghats of river Ganga at Haridwar. National Conference on status of Indian Environment, Pre-conference proceeding, ASEA. Rishikesh, p-38.

Khanna, D.R., Bhutiani, R. and Trivedi, Manoj (2002). Impact of paper mill effluent on some water parameters of Hindon river, at Sharanpur. *Him. J. Env. Zool., Vol.* 16 (1) , pp 125-128.

Khanna, D.R. and Bhutiani, R. (2003). Limnological characteristics of river Ganga at Haridwar. *Uttar Pradesh J. Zool.*, 23(3) :179-183.

Khanna, D.R. and Bhutiani, R. (2003). Limnological status of Satikund pond at Haridwar *Indian J. Env. Sciences* , 7 (2), pp. 131-136.

Khanna, D.R. and Singh, R.K. (2000). Seasonal fluctuations in the plankton of Suswa River at Raiwala (Dehradun). *Env. Cons. J.* (2 & 3) : 89-92.

Khanna, D.R. and Singh, S. (2002). *Pond fish Ecology and Economics.* Daya Pub. House, N.Delhi. pp.1-81.

Khanna, D.R., Seth, T.R., Malik, D.S. and Singh, R.K. (1999). Fish and their ecology of river Suswa at Raiwala, Dehradun. *Proc. Nat. Conf. in Sus. Eco. and Env.*, pp. 105-110.

Khanna, D.R., Singh Shakun, Gautam, A. and Singh, J.P. (2003). Assessment of water quality of river Ganga in District-Bulandshahar (U.P.) *India. J. Nat. Conservation*, 15(1) : 167-165.

Khanna, D.R. and Bhutiani, R. (2004). Fish and their ecology of river Ganga at Gohri ghat, Garhwal (Uttranchal). In: *Fish diversity in protected habitat.* pp.291-298

Kumar, A. (1995). Studies on pollution in river Mayurakshi in South Bihar. *Indian J. Env. Poll.,* 2 (1) 21-26.

Kumar, A. and Bohra, R. (1999). Gastropods as indicators of pollution status of some wetlands in Santal pargana, Bihar, India. *Indian J. Env. and Ecoplan.,* 2(1): 83-87.

Kumar, S. Sathya and Mukherjee, S.K. (2000). *Recent changes in the flora and fauna of Doon Valley and surrounding area.* A report submitted to the Supreme Court Monitoring Committee. W.I.I., March 2000, pp. 27-45.

Kumar, Neeraj and Sharma, R.C.(2002). Water quality of Krishni river (Physico-chemical parameters*). J. Natcon,* 14(2) 273-297.

Kumar, Ravinder (2002). Influence of ammonia to fresh water murrel *Channa punctatus* (Bloch). *Nature Environment and Pollution Technology*, 1(4): pp. 365-367.

Kumar, Amod, Naushad, Alam and Pandey, B.N. (2002). Performed studies on the effect of metacid stress on toxicity and behavioral responses in *Channa puncatus* (Bloch). *Him. J. Env. Zool.,* Vol.9, pp. 61-62.

Kuhen, I., Allestam, G., Engdahl, M., Stenstrom, T.A., Moris, R., Grabow, W.O.K. and Jofer, J. (1997). Biochemical finger printing of Coliform bacterial populations- Comparison between polluted river water and factory effluents. *Health related Water Microbiology*. pp. 343-350.

Kudesia , V.P. and Verma , S.P. (1985). Industrial pollution on Kali river *J. Env. Sci.,* 1 (2) : 41-49.

Lal, A.K. (1996). Effects of mass bathing on water quality of Pushkar Sarpvar. *Indian J. Environ. Prot*, 16 (11). Pp. 831-836.

Lakshminarayanan, J.S.S. (1965). Studies on phytoplankton of the river Ganges, Varanasi, India. *Hydrobiological*, 25 : 119-164.

Leidy, R.A. and Moyle, P.B. (1997). Conservation status of world fish faunas: an over view. In Conservation Biology for the coming decade (Eds. Fielder, P.L. and Kareiva, P.M.) *Chapman and Hall, London*, pp. 187-227.

Lloyd, R. (1961). Study on the toxicity of mixture of zinc and copper sulfates to rainbow trout (*Salmo gaidnerii*). Richardson. *Ann. App.Biol.,* 49: 535-538.

Macan, T.T. (1974). *Fresh Water Ecology*, 2nd Edition, New York: Wiley, 224 pp.

Mahajan, C.L. (1961). Fish fauna of Muzzafarnagar district Uttar Pradesh. *J. Bombay Nat. Hist. Soc.,* 62 (3): 440-454.

Mahatre, G.N., Chaphekar, S.B., Ramani Rao I.V., Patil, M.R. and Haldhar, B.C.(1980). Effect of industrial pollution on the Kalu river ecosystem. *Environ. Poll. Series,* A 23: 67-78.

Mazhar, M.D. and Kapoor, C.P. (1992). Limnological study on Doraniu river at Bareilly. (U.P.) *J. Fresh Water Bio.* 4 (2) : 155-158.

Malik, D.S., Bhanot, M and K.S. Negi (2003). Impact of distillery effluent on the water quality and phytoplankton in river Saung. *Indian J. Env. Sciences*, 7 (2), pp. 163-167.

Mills, H.B., Starrett, W.C. and Bellrose, F.C. (1966). Man's effect on the fish of river Illinois. *Nat. Hist. Surv. Biol.* Notes. 57.

Mittal, S. and Sengar, R.M.S. (1990). Phytoplanktonic diversity in relation to certain physico-chemical characteristics of river water. *Proc. Nat. Acad. Sci.* India. 60(B), III.

Mishra, N. and Saxena, A.B. (1984). The effect of sewage with special reference to aquatic insects in the river Kshipra (India). *Int. J. Environ. Studies,* 23 : 191-208.

Mishra, Sarika and Joshi, Namita (2002). Generation of solid waste during festive occasions in Haridwar city. *Him. J. Env. Zool., Vol.* 16 (2), pp 263-266.

Mishra, S. and Joshi, B.D. (2003). Assessment of water quality with few selected parameters of River Ganga at Haridwar. *Him. J. Env. Zool.* Vol. 17(2) : 113-122.

Mohanty, R.C. (1981). Water quality of some water bodies of Bhubaneshwar, *PhD. Thesis, Utkal University*, p1-240.

Monichan, K.K. and Sharma, R.K. (2003). Distribution status of fish fauna in Etawah district, U.P. *Indian J. Env. Sciences,* 7(2), 115-117.

Motwani, M.P., Banerji, S.M. and Karam Chandani, S.J. (1956). Some observation on the pollution of river Sone by the factory effluent of Rohtas industries at Dalmia nagar (Bihar). *Indian J. Fish.,* 3:334-367.

Moyle, P.B. and Williams, J.E. (1990). Biodiversity loss in temperate zone: Decline of natural fish fauna of California. *Conservation Biology*: 4: 474-484.

Muller, G.(1966). *Die sedimentolldes in Bodensee.* Naturuissenschaften 53 (10) : 237-247.

Muller, G. and Forstner, U. (1968 b). General relationship between suspended sediment concentration and water discharge in Alpien Rhein and some other rivers. *Nature* 217 : 244-245.

Murugesan, A.G., Abdul Hameed, K.M.S.A. and Sukumaran, N. (1994). Water quality profile of the perennial river Tampraparni. *Indian J. Env. Prot.,* Vol.14 (8) 567-572.

Nautiyal, P. (1990). Ecology of the Ganga river system in the upland of Himalaya. *Environment. Resource and Development*, 69-73.

Nautiyal, P. and Lal, M.S. (1982 a). Fecundity of Garhwal Himalayan Mahseer, *Tor putitora* (Ham.) *J. Bombay Nat. Hist. Soc.,* 82: 253-257.

Nautiyal , P. and Lal, M.S. (1982 b). Recent record of Garhwal Himalayan Mahseer (*Tor putitora*) with a note on its present status. *J. Bombay Nat. Hist. Soc.* , 79: 639-695.

Nautiyal, P. , Dobriyal, A.K., Chaukiyal, D.C., and Singh, H.R. (1986). Quality and potability of Alaknanda river at Srinagar- Garhwal

(U.P., India) A note on Bacterial density as an indicator of pollution. *Poll. Res.,* 5(3-4) : 143-145.

Nautiyal, P. (1990). Ecology of the Ganga river system in the upland of Garhwal Himalaya. *Environment Resource and Development:* 69-73.

Nautiyal, P. and Singh, H.R. (1989). River projects and the endangered Himalayan *Tor putitora* in the uplands of Garhwal region. A critical evaluation. *Him. J. Env. Zool.*, 3: 129-134.

Needham, J.G. and Needham, P.R. (1972). *A guide to the study of Fresh Water Biology*. Holden-Day 1. N.C.San Francisco, *Coilt.* 94 (3): 1-108.

Oliff, W.O. (1959). Hydrobiology of Tugela river system. Part I. The main Tugela river *Hydrobiologia*, 14: 281-370.

Pahwa, D.V. and Mehrotra, S.M. (1966). Observations on fluctuations in abundance of plankton in relation to certain hydrological conditions of river Ganga. Proc. Nat. Acad. Sci. 36B (2): pp157-158.

Pandey, B.N., Kumar. K and Das, P.K.L. (1993). A priliminary study on the physico-chemical quality of water or river Kosi. *J. Ecobiol.*, 5: 237-339.

Pandey, K.P., Kumari, V. and Pandey, A.K.(1996). Study of diurnal variation in the physico-chemical parameters of two lentic water bodies of Darbhanga. *Him. J. Env. Zool.,* Vol. 10,pp 61-62.

Pandey, S., Kaushik, P. and Parashar, N.B. (2003). Physico-chemical and Microbiological Studies of Ganga Canal at Haridwar, During Kumbh Period-1998. *Him. J. Env. Zool.* Vol. 17(2), 167-171.

Pant, M.C. (1970). Fish Fauna of Kumaon hills. *Rec. Zoo. Surv. India*, 64 (1-4) :85-96.

Patrick, R. (1978). Effect of trace metals in the aquatic ecosystem. *Am. Scientist,* 66 (2).

Patil, M.R.and Patil, A.R. (1983). Water quality of the Ulhas river with reference to Copper, Cadmium and Mercury. *Poll. Res.* 2(1): 24-27.

Pathak, J.K. (1991). Some aspects of ecology of the rivers Sarju and Gomti of the Kumaun Himalaya. *Ph.D. Thesis. Kumaun University Nainital.*

Pathani, S.S., Upadhya, K.K. and Joshi, S.K. (2002). Some physico-chemical parameters and primary productivity of river west Ramganga. *Him. J. Env. Zool.,* Vol. 16(2), pp. 151-158.

Pokhriyal, R.C., Dobriyal, A.K. and Singh, H.R. (1983). Seasonal variation in potamological parameters of the stream Khandagadh. *Uttar Pradesh J. Zoo.*, 3:70-72.

Prescott, G.W. (1972). The Algae. A Review. *Riverside studies in Biology*. 41, 280 pp.

Prasad, B.N. and Yashpal, S. (1982). On diatoms as indicators of water pollution. *J. Indian Bot. Society*, 61: 326-336.

Prasad, Guru B. (2003). Physico-chemical and bacteriological quality of ground waters at Tadepalli Mandal of Guntur district, Andhra Pradesh. *Nature Environment and Pollution Technology,* Vol.2, No.2., pp. 173-178.

Prasad, B.B. and Singh, R.B. (2003). Composition, abundance and distribution of phytoplankton and zoobenthos in tropical water body., *J. Not. Env. and Poll. Tech.,* Vol. 2 (3):255-258.

Prasad G., Shankar V., and Chopra A.K. (2003). Environmental state of River gangas in Rishikesh Haridwar and its management. *Env. Cons. J.* 4 (1-3) 51-58.

Raina, V., Saha, A.R. and Ahned, S.R. (1984). Pollution studies on river Jhelum-1: Assessment of water quality. *Indian J. Env. Health,* 26: 187-210.

Raina, R., Subla A.R. and Zutshi, D.P. (1982). Water quality and plankton of Jhelum River. *Int. J. Ecol. Environ. Sci.* 8 : 11-17.

Raj, D.Selva, Thampi Jeyaraj, C. and Chithra, G.(2003). Made studies on hydrology and fish species diversity of Thamiraparani river in Kanyakumari district. *Nature Environment and Pollution Technology,* Vol.2, No.3, 323-326.

Ramesh, N.B., Srikanth, R. Murlimohan Rao, A. (1992). Evaluation of water quality in three rivers of Andhra Pradesh. *Poll. Res.*, 11 (4) : 209-212.

Rao, K.L. (1975). *India's Water Wealth*, Orient Longman, N.Delhi. pp.255.

Ray, H.K. (1955). Plankton ecology of the river Hoogly at Palta. *Ecology*, 36(2), 169-175.

Ray, P., Singh, S.B. and Sehgal, K.L. (1966). A study of some aspects of the river Ganga and Jamuna at Allahabad (U.P.) in 1958-59. *Proc. Nat. Acad. Sci. India,* 36 B (3): 235-272.

Reddy, P.M. and Venkateswarlu, V. (1987). Assessment of water quality and pollution of river Tungabhadra. Near Kurnool, (A.P.) *J. Environ. Biol.,* 8 (2) : 109-119.

Reid, K.G. (1961). Ecology of Inland waters and Estuaries New York: Reinhold, 280 pp.

Robinson, J.M. and Kallar, E.C. (1976). A comparison of water characteristics of four north wastern Verginian rivers. *Jour. Proc.w.va. Acad. Sci.*, 48 (1): 1-67.

Sangu, R.P.S. and Sharma, K.D.(1985). Studies on water pollution on Yamuna river in Agra. *Indian J.Environ. Health.*, 27(3): 257-261.

Saxena, K.K. and Chauhan, R.R.S. (1993). Physico-chemical aspects of pollution in river Yamuna at *Agra. Poll.Res.* 12(2):101-104

Sehgal, J.K. (1973). Fish and fisheries of Aligarh with notes on fish catche. *J.Inland Fish Soc. India*, Vol. V:23-28.

Sengar, R.M.S., Sharma, K.D. and Pathak, P.D. (1985). Studies on distribution of algal flora in polluted and non-polluted regions in Yamuna river at Agra (U.P.). *J. Indian Bot. Soc.* 64 : 365-376.

Shanthi, K., Ramasamy, K. and Lakshmanaperumalsamy, P. (2001). Hydrobiological study of Singanallur lake at Coimbatore, India. *Nature Environment and Pollution Technology,* 1 (2), pp.97-101.

Sharma, A. (1999). Limnological studies of Ban Ganga and distributional pattern of stream bottom fauna. *Ph. D. Thesis*, University of Jammu.

Sharma, R.C. (1986). Effect of Physico-chemical factors on benthic fauna of Bhagirathi river Garhwal Himalaya , *Indian J. Ecology*, 13(1): 133-137.

Sharma, V.K. (1986). The biology and fishery of *Cyprinus carpio* (Lin.) from Gobind Reservoir, H.P. India, *Ph.D. Thesis*, Punjab University Chandigarh.

Sharma and Pandey (1998). Pollution status on Ramganga river at Moradabad*Poll. Res.* 17 (2) : 201-209.

Sharma, S.K., Gurutu, J.N. and Verma, S.R. (2000). Characteristics and pollutional effects of pulp and paper mill waste on Tapti river, District Surat. *J.Natcon.* 12(2) : 293-306.

Sharma, S.K., Mudgal, L.K. and Balsare, D.K. (2002). Biodiversity benthic communities of Sirpur Lake, Indore. *Him. J. Env. Zool.,* Vol. 16 (2), pp. 231-232.

Seth, T.R., Khanna , D.R., Gautam, A., Chugh, T. and Sarkar, P. (2000). Temporal trends of phytoplanktonic diversity in the river Ganga at Haridwar. *Him. J. Env. Zool.* Vol. 14, 129-134.

Shivsubramani , R. and Mahadevan, A. (1995). Water quality of river (Suruliyar) in Tamil Nadu *Poll. Res.*, 14 (1) : 72-82.

Shrivastava, V.K., Shrivastava, G.K. and Shrivastava, J.K. (1996). Phytoplankton productivity and physico-chemical properties of Rapti river *Ecol. Env. and Conc,* 2 (184-185)

Shukla, Arvind and Srivastava, Sharad (2003). Observations on macrozoobenthos saprobiety index with relation to pollution status of Gandhisagar reservoir in Madhya Pradesh. *Uttar Pradesh J. Zool.* 23, pp.185-190.

Singh, H.R. (1991*). Advances in limnology*. Narendra Pub. House, N.Delhi, p. 322.

Singh, H.P, Choudhry, M. and Kalekaur. V. (1982). Seasonal and diurnal changes in physico-chemical features of the river Brahmaputra at Gauhati, *Ind.J.Zool.,* 2(82):77-84.

Singh, HR., Badola, S.P. and Dobriyal, A.K. (1983). Geographical distributional list of ichthyofauna of the Garhwal Himalaya with some new records *J. Bombay Nat. Hist. Soc.*, Vol. 84 (1) : 126-132.

Singh, J.P., Yadava, P.K. and Singh, L. (1988). Pollution status on Sangam and its adjoining river before the Kumbh Mela at Allahabad. *Indian J. Environmental Protection.* Vol. 8(11) : 839-842.

Singh, J.P., Yadava, P.K. and Singh, L. (1989). The assessment of water quality of Sangam and its adjoining river Ganga and Yamuna after Maha Kumbh Mela at Allahabad. *Indian J. Environmental Protection.* 1 (5) : 372-375.

Singh, J.P., Upadhyay, R. and Singh, Shakun (1994). A study of water quality of river Yamuna at Agra with special reference to coliform bacteria. *Res. J.P.I. Envi.* 10: 53-57.

Singh, B.K., Singh, S.K. and Singh, B.B. (2002). Distribution and seasonal abundance of benthic macro invertebrates in the Bisar sarovar of Gaya, Bihar. *Him.J.Env.Zool.,* Vol. 16(2), pp. 139-150.

Sivakumar, A.A., Arunadevi, P and Aruchami, M. (2003). Studied the water quality of the river Ambarampalayam, Coimbatore District, Tamilnadu. *Nature Environment and Pollution Technology,* Vol. 2, No. 3, pp. 305-308.

Sivakumar, A.A., Thirumathal, K. and Aruchami, M. (1987). Effect of pollution on the physico-chemical properties of the river Amravati (Tamil Nadu). *Proc. Nat. Conf. Environ. Impact Bio.,* Madras.

Somashekhar, R.K. (1984). Phytoplankton as indicators of water quality. A study of river Cauvery. *Int. J. Environ. Studies*, 24: 115-123.

Strommer, J.L. and Smock, L.A. (1989). Vertical distribution and adundance of invertebrates within the sandy substrate of low gradient head water stream. *Fresh Water Biology*, 22 : 263-274.

Stumm, W. and Bilinski, H. (1973). Trace metals in natural waters difficulties of interpretation arising from our ignorance of their speciation in "Advanccs in Water Pollution Research". *Proceedings 6th International Conference* (ed. S.H. Jekin) 946: 39-52.

Sukumaran, P.K. (2002). Coliform bacteria an index of aquatic pollution. *Trends in life Sciences* (India), 17(2), 19-24.

Tan, T.H. (1971). On occurrence and ecological adaptation of zoological plankton in Tanshi river, Taiwan. *Rep. Inst. Fish. Biol. Taiwan*, 2(4):40-48.

Talling, J.F. and Rzoska, J. (1967). The development of plankton in relation to hydrobiological regime in Blue Nile. *J. Ecol.* 55 : 636-662.

Tombi, Singh, H. (1992). Ecobiology of hill stream fishes of Manipur with special reference to culturable possibilities. *Final tech. report. DSTE, Govt. of Manipur*, pp. 1-50.

Train, R.E. (1967). In: *Quality criteria of water* US Environmental protection agency, Washington.

Trivedi, R.K. and Goel, P.K. (1984). *Chemical and Biological method for water pollution studies*. Karad, *Environmental Publication*, 1-25.

Velz, C.J. (1947). Factory influencing self purification and their relation to pollution abatement (I) *Sewage work J.,* 19:623-644.

Venkateshwarlu, T. and Jayanti, T.V. (1968). Hydrobiological studies of the river Sabarmati to evaluate water quality. *Hydrobiologia,* 31: 442-448.

Venkateshvwarlu, T. and Menon, A.G.K. (1979). A list of fishes of river Gangas and its branches. *Hydrobiologia*, 9: 46-70.

Verma, S.R. (1978). Pollution studies of few rivers of western Uttar Pradesh with reference to biological indices. *Proc. Indian Acad. Sci.,* 87 B: 121-131.

Verma, S.R. and Dalela, R.C. (1975a). Studies on the pollution of Kalinadi by industrial waste in Mansurpur. Part I. Hydrometric and physico-chemical characteristic of waste and river water. *Acta. Hydrochim. Et. Hydrobiol*, 3: 239-257.

Verma, S.R., Sharma. P., Tyagi,A., Rani, S., Gupta, A.K. and Dalela, R.C. (1984). Pollution and saprobic status of eastern Kalinadi. *Limnologica,* 15 (1): 69-133.

Verma, S.R. and Shukla, G.R. (1969). Pollution in perennial stream Khala, by the sugar factory effluent near Laskar (Dist. Saharanpur), U.P., *India, Env. Health*, 11: 145-162.

Ward, R.W. and Seibert, H.C. (1963). Investigation into the limnology of Dow lake, Athenes Country. *Ohio J. Sci.,* 63(3):128-144.

Ward, H.B. and Whipple, G.C. (1959). *Fresh Water Biology* 2nd edition. John Wily and Sons. INC. N.York, London.

Weible, S.R., Anderson, R.L.Woodward, R.L. (1964). Urban land run off as a factor in stream pollution. *THIS.J.*36 (7): 914.

WHO (1985). *Guide lines for drinking water quality* 1st edtion, pp.30.

Zingde, M.D., Narvekar, P.V., Sharma, R.V. and Desai, B.N. (1980). Water quality of the river Damanganga (Gujarat). *Indian. J. Mar. Sci.*, 9 : 94-99.

INDEX

F

G

H

I

X

Y

Z